Selectors

Selectors

John E. Jayne

and

C. Ambrose Rogers

PRINCETON UNIVERSITY PRESS
PRINCETON AND OXFORD

Published by Princeton University Press,
41 William Street, Princeton, New Jersey 08540

In the United Kingdom: Princeton University Press,
3 Market Place, Woodstock, Oxfordshire OX20 1SY

Library of Congress Cataloging-in-Publication Data applied for.
Jayne, John E. and Rogers, Ambrose C.
Selectors / John E. Jayne and C. Ambrose Rogers
p. cm.
Includes bibliographical references and index.
ISBN 0-691-09628-7 (alk. paper)

British Library Cataloguing-in-Publication Data
A catalogue record for this book is available from the British Library.
This book has been composed in Times and Abadi

Printed on acid-free paper. ∞

www.pup.princeton.edu

Printed in the United States of America

10 9 8 7 6 5 4 3 2 1

Contents

Preface

Suppose that $F(x)$ is a non-empty set in a space Y for all x in a metric space X. A point-valued function s from X to Y that satisfies $s(x) \in F(x)$ for all $x \in X$ is called a selector for F. Two problems have been studied in many forms by many authors:

(a)   When can a continuous selector be found for F?

(b)   Given a measure on X, when can one find a measurable selector for F?

The theory of continuous selectors for lower semi-continuous set-valued maps, after study by several authors, has been brought to a very satisfactory state by E. Michael, see Chapter 1. The simplest of upper semi-continuous set-valued map are easily seen to lack continuous selectors. For example, if we define $f : [0, 1] \rightarrow \mathbb{R}^2$ by

$$F(x) = 0, \text{ if } 0 \leq x < 1/2,$$
$$F(x) = 1, \text{ if } 1/2 < x \leq 1, \text{ and}$$
$$F(1/2) = \{(1/2, y) : 0 \leq y \leq 1\},$$

then F is an upper semi-continuous map on $[0, 1]$, whose values are non-empty compact convex subsets of $\mathbb{R}^2$, and F clearly does not have a continuous selector.

For sometime we have thought that the theory of measurable selectors was somewhat lacking, in that one knows little about the topological properties of measurable functions. It is surprising that in very general circumstances upper semi-continuous set-valued maps do have selectors that, although not continuous, are of the first Baire class; that is, are the pointwise limits of sequences of continuous functions.

In the book we are mainly concerned with proving results showing the existence of selections of the first Baire class. We give a number of geometrically interesting examples, and some unexpected consequences for functional analysis.

J. E. Jayne
C. A. Rogers

Introduction

Zermelo [81] presented a proof that any set can be wellordered. His proof was based on the apparently clear idea that if M is any set with a given cardinal number, then it is possible for each nonempty subset M' of M to choose an element m' of M' and call it the distinguished element of M'.

Borel [4] wrote accepting that Zermelo had proved that the proposition: "A set M can be expressed as a wellordered set" was implied by the proposition: "Given an arbitrary nonempty subset M' of a set M it is possible to choose a point m' of M' and call it the distinguished element of M'."

He also remarked that the second proposition follows from the first.

However, Borel did not accept the second proposition and felt that an argument that exhausts the elements of an uncountable set by use of a transfinite sequence of choices was outside the domain of mathematics.

This dispute led to much discussion and in particular to letters from Hadamard to Borel, Baire to Hadamard, Lebesgue to Borel and Borel to Hadamard [18] or see the reprint in Hadamard [17].

The general consensus seems to have been that these were two acceptable types of mathematicians, each with their own type of mathematics. The "realists" felt that mathematics should be concerned with entities that could be defined in a finite number of words or entities that could be established as the unique realization of a specification written in a finite number of words. On the other hand, the "idealists" could use their imagination freely and could imagine the completion of transfinite operations. This foreshadowed the distinction between those mathematicians who refuse to use the axiom of choice and those who use it freely.

Lusin in his book [51] discusses these issues in depth.

In this book we study selection theorems that are related to the axiom of choice. Indeed, the most general form of the selection problem is equivalent to the axiom of choice.

First selection problem *Let F be a set-valued map from a set X to a set Y, so that $F(x)$ is a nonempty subset Y for each x in X. Is there a point-valued map f, called a selector for F, from X to Y with $f(x) \in F(x)$ for each x in X?*

Since this selection problem is clearly equivalent to the axiom of choice, we are concerned with a less well-defined problem that we shall study in many special cases.

Second selection problem *Let F be a set-valued map with nonempty values from a topological space X to a topological space Y. When can one find a point-valued function f from X to Y with f(x) ∈ F(x) for all x in X and arrange that f is a function of some given Borel or Baire class?*

To ensure that the answer to this second problem is "Yes!" we need to impose quite strong conditions on the spaces X and Y and on the set-valued function F; the class of the selector f will depend on the conditions imposed. In most of the results we need to use the axiom of choice, but in a subtle, indirect way; one cannot expect to find a selector that belongs to any Borel or Baire class by a direct use of the axiom of choice.

When the problem is formulated in very specific terms the selector may well be obtained without appeal to the axiom of choice. Consider the following simple example. Suppose that X and Y are both the unit interval and that the set-values of F are nonempty intervals of $[0, 1]$ and that

$$G(F) = \bigcup \{x \times F(x) : x \in X\}$$

is a closed convex set in $[0, 1] \times [0, 1]$. In this case the selector f defined by

$$f(x) = \inf\{y : y \in F(x)\}, \quad \text{for } 0 \leq x \leq 1,$$

is a continuous selector for F.

Perhaps the first nontrivial example was obtained by P. S. Novikov in 1929. His result was stated by N. Lusin in 1929 (see [50, the footnote on page 315]). Novikov published his proof in 1931 [63].

Theorem (Novikov) *Let F be a map from ℝ to ℝ taking only nonempty countable values, the graph*

$$G(F) = \bigcup \{(x, y) : x \in \mathbb{R}, \ y \in F(x)\}$$

being a Borel set in $\mathbb{R}^2$. Then F has a selector f whose graph in $\mathbb{R}^2$ is also a Borel set.

Novikov's paper is best read in conjunction with Lusin's 1924 paper [49]. Lusin shows, by general arguments that any Borel set B in $\mathbb{R}^n$ will be finite or countable or will be a set of values of a function $f : \mathbb{R} \to \mathbb{R}^n$ that is continuous at all but countably many points of $\mathbb{R}$ and that never takes the same value in $\mathbb{R}^n$ twice. Naturally the same uncountable Borel set in $\mathbb{R}^n$ will have many representations in this form. Lusin supposes that such a representation has been chosen, but recognizes that there is no general way in which this choice can be made constructively. Once the choice has been made, Lusin constructed a G_δ-set in $\mathbb{R}^{n+1}$ of a rather special type that coincides with the graph of f, and so yields the Borel set B as the injective projection of the G_δ-set from $\mathbb{R}^{n+1}$ into $\mathbb{R}^n$.

Novikov starts from this type of representation of the graph G of his set-valued function F, taking nonempty countable values, as the injective projection on $\mathbb{R}^2$ of a G_δ-set in $\mathbb{R}^3$. Using general arguments, involving the use of all countable ordinals, Novikov establishes a number of special properties of the sets he is considering, establishing such a well-controlled situation that he is able to obtain his selector f constructively and to show that its graph is a Borel set.

In the second part of his paper, Novikov gives an example to show that one may not omit the hypothesis that the set-valued function F takes only countable values.

Since we want to obtain selectors of small Borel or Baire classes, we do not give any proof of Novikov's result; nor do we describe the many results of a similar nature that have been obtained by a succession of authors. See W. Sierpiński [71], S. Braun [5], M. Kondô [46], M. Sion [73], Y. Sampei [70], Y. Suzuki [79], and C. A. Rogers and R. C. Willmott [67]. For a survey of more recent work, see D. A. Martin and A. S. Kechris [53].

When one investigates more abstract situations, it seems that to obtain satisfactory results one needs to be able to use the axiom of choice and other nonconstructive methods to impose a sufficiency of structure on the initial situation, and perhaps also on situations that arise in the course of the proof, in a way that enables the main line of proof to proceed constructively.

We illustrate this in relation to Michael's continuous selection theorem, Theorem 1.1 below. The proof, as we have presented it, appears, at first sight, to be constructive. However, the existence of the partitions of unity that are employed in the proof require justification by use of the axiom of choice to impose a sufficiency of structure on the spaces and on the lower semi-continuous set-valued maps that are used. We discuss this in some detail in Remark 11 at the end of Chapter 1.

Although similar considerations arise in the other selection problems that we consider, we do not pursue the matter further.

We now give an outline of the main results discussed in this book. Let F be a set-valued map from a topological space X to a topological space Y, so that $F(x)$ is a set in Y for each x in X. The set-valued map F is said to be *upper semi-continuous* if

$$F^{-1}(C) = \{x \in X : F(x) \cap C \neq \emptyset\}$$

is closed in X for each closed set C in Y; it is said to be *lower semi-continuous* if $F^{-1}(U)$ is open in X for each open set U in Y; and it is said to be of the *first lower Borel class* if $F^{-1}(U)$ is an $\mathcal{F}_\sigma$-set in X whenever U is an open set in Y.

Note that, if the open sets in Y are $\mathcal{F}_\sigma$-sets, then each upper semi-continuous F is of the first lower Borel class. A point-valued function f from X to Y is said to be a *selector* for F if $f(x) \in F(x)$ for each x in X.

A point-valued function $f : X \to Y$ is said to be of the *first Borel class* if $f^{-1}(U)$ is an $\mathcal{F}_\sigma$-set in X whenever U is open in Y; it is said to be of the *first Baire class* if it is the pointwise limit of a sequence of continuous functions from X to Y. If Y is a metric space, each function of the first Baire class is automatically of the first Borel class.

E. Michael [54] has shown that a lower semi-continuous map, from a paracompact space X to a Banach space Y, taking only nonempty closed convex values, has a continuous selection; he goes on to prove that X is necessarily paracompact if this holds for every Banach space Y. In the first part of Chapter 1 we give an account of Michael's proof of the existence of his continuous selection when X is a metric space.

K. Kuratowski and C. Ryll-Nardzewski [48] have shown that a set-valued map F of the first lower Borel class from an arbitrary metric space X to a *separable* metric space Y, with nonempty complete values, has a selector of the first Borel class. Indeed, they give a very general abstract version of this result; but their method seems to depend essentially on the separability of the metric space Y. We give an account of this work in the second part of Chapter 1.

The selectors that we construct in Chapters 3–7 are obtained as limits of functions that are constant on the sets of certain partitions of the space X. It is economical to devote Chapter 2 to the study of functions of this type. If readers tackle Chapter 2 seriously before looking at the subsequent sections, they may well find this section very dry. We suggest that they skip quickly through Chapter 2 first noting the main definitions and Theorem 2.1, and then return to this section, when they find that they need to, in the course of studying Chapter 3 onwards.

A Banach space Y is said to have the *Point of Continuity Property* if the identity map on Y restricted to any bounded weakly closed set has a point of weak-to-norm continuity. R. W. Hansell, J. E. Jayne and M. Talagrand [22] prove, in particular, that *an upper semi-continuous set-valued map from a metric space X to a Banach space Y, having the Point of Continuity Property, taking only nonempty weakly compact values, has a selector that is of the first Baire class as a map from X to Y with the norm topology.* We give an account of their work in Chapters 2 and 3 establishing some of their refinements.

In a recent paper N. Ghoussoub, B. Maurey and W. Schachermayer [14] give a number of selection results and applications. A simplified version of one of their main results takes the following form. *If Y is a Banach space with the Point of Continuity Property there is a map s from the space of nonempty weakly compact sets of Y to Y such that:*

(1) $s(K) \in K$ *for each nonempty weakly compact K;*

(2) *if $\emptyset \neq K' \subset K$ and $s(K) \in K'$, then $s(K) = s(K')$;*

(3) *if $\{K_\alpha : \alpha \in D\}$ is a decreasing net of nonempty weakly compact sets in Y then*

$$\left\| s(K_\alpha) - s\left(\bigcap_{\alpha \in D} K_\alpha \right) \right\| \to 0$$

as $\alpha \to \infty$ through D;

(4) *if M is a metric space and $F : M \to (Y, weak)$ is an upper semi-continuous set-valued map, taking only nonempty compact values, then $s \circ F$ is a selector for F that is of the first class when regarded as a map from M to Y with its norm topology.*

Note that each weakly compact set in Y is a bounded norm closed set in Y, so that the space of nonempty weakly compact sets of Y has a natural metric, the restriction of the Hausdorff metric on the bounded norm closed sets of Y. In Chapter 4 we give an account of this result of Ghoussoub, Maurey and Schachermayer, showing also that s is of the first Baire class when regarded as a map from the space of nonempty weakly compact sets with the natural metric to Y with its norm. We do not attempt to summarize the rest of the paper by Ghoussoub, Maurey and Schachermayer; we commend this paper to the reader.

In Chapter 5 we give some applications of the theorems that we have already proved. We develop the theory of maximal monotone maps and obtain a selection theorem for such maps. This leads directly to a selection theorem for the subdifferential of a continuous convex function. We also obtain selection theorems for some geometrically defined maps: attainment maps and nearest point maps. We also discuss a rather different type of problem. Suppose that a linear programming problem with a bounded feasible region in $\mathbb{R}^n$ depends on a point p in a parameter space P. Suppose further that the feasible region $C(p)$ and the (linear) objective function depend continuously on the parameter point p. In these circumstances one cannot expect to find optimal solutions, that depend continuously on p. However, we show, under appropriate conditions, that there will be a sequence of continuous maps $f_n :$ $P \to \mathbb{R}^n$ with $f_n(p) \in C(p)$ for all p in P and such that, for each p in P, the sequence $f_n(p)$ converges to an optimal solution of the problem corresponding to p.

V. V. Srivatsa in 1985–86 discovered and wrote up some striking results; they were published much later in [74]. We quote two of his results here.

An upper semi-continuous set-valued map from one metric space to another, taking only nonempty values, has a selector of the first Borel class.

An upper semi-continuous set-valued map from a metric space to a Banach space with its weak topology, taking only nonempty values, has a selector that is of the first class when regarded as a map from the metric space to the Banach space with its weak topology.

We give an account of these and other results in Chapter 6.

In Chapter 7, we give two further applications of the theory of selections.

M. Fabian and G. Godefroy [10] have shown that *the dual of every Asplund space has an equivalent locally uniformly convex norm.* This they do despite the existence of an example of an Asplund space that has no equivalent norm that has a locally uniformly convex dual norm. We only give an outline of their proof, since, although we do describe the theory of Asplund spaces (see [61]), to give the proofs of the theory would take us too far from the main theme of this book. Further, for the same reason, we quote, without any discussion, a result from renorming theory.

The second part of Theorem 5.4 below has a converse: *Let K^* be a non-empty weak* compact set in the dual X^* of a Banach space X; if the attainment map from X to K^* has a selector f that is of the first class as a map from $(X, norm)$ to $(X^*\ norm)$, then K^* is a weak* fragmented by the norm in X^*.* Examples show that this converse does hold *provided X contains no isomorphic copy of $\ell_1(\mathbb{N})$.* We follow quite closely the attempt of Jayne, Orihuela, Pallarés and Vera [41] to prove the converse without any proviso. We again need to quote results from the theories of Asplund Spaces and of Renorming.

In Chapter 7 we can only give proofs that take for granted some difficult specialized results that we have to quote without proof.

Selectors

Chapter 1

Classical results

Many selection theorems have been proved; many of the earlier ones are called uniformization theorems. One of the important starting points is the work of N. N. Lusin and P. S. Novikov [52]. However, the first general results yielding continuous selectors and selectors of the first Borel class were obtained by E. Michael [54] and by K. Kuratowski and C. Ryll-Nardzewski [48]. We take these two results as our starting points.

1.1 MICHAEL'S CONTINUOUS SELECTION THEOREM

Although Michael's work was published in three papers in the *Annals of Mathematics* [54,55,56], we shall follow his paper in [57], confining our attention to the special case when the domain is a metric space.

Theorem 1.1 (Michael)   *Let F be a lower semi-continuous set-valued function from a metric space X to a Banach space Y, and suppose that F takes only nonempty closed convex values in Y. Then F has a continuous selector.*

Following Michael [54] we obtain the following theorem as a corollary.

Theorem 1.2 (Bartle-Graves)   *If Y and X are Banach spaces and u is a continuous linear transformation from Y onto X, then there is a homeomorphism f mapping X to a subset of Y such that*

$$f(x) \in u^{-1}(x)$$

for every x in X.

We shall need a standard, but nontrivial result from the theory of metric spaces. If $\mathcal{U}$ is any open cover of a metric space X, a family

$$\{p_\gamma : \gamma \in \Gamma\}$$

of continuous functions from X to the unit interval is said to be a partition of unity on X refining the family $\mathcal{U}$ if:

(a) each x in X has a neighborhood N_x with p_γ not identically zero on N_x for only a finite set of γ in Γ;

(b) for each x in X

$$\sum \{p_\gamma(x) : \gamma \in \Gamma\} = 1$$

and

(c) for each γ in Γ there is a U_γ in $\mathcal{U}$ with $p_\gamma(x) = 0$ on $X \setminus U_\gamma$.

Lemma 1.1 *Let $\mathcal{U}$ be an open cover of a metric space X. Then there is a partition of unity on X refining the family $\mathcal{U}$.*

The proof of this lemma may be found, for example, in Engelking [8, p. 374]. If S is any set in a Banach space Y and $r > 0$, we use

$$B(S; r) = \{y : \|y - s\| < r \text{ for some } s \text{ in } S\}$$

to denote the expansion of the set S by r. Clearly $B(S; r)$ is always open in Y, and it is convex when S is convex.

Lemma 1.2 *Let F be a lower semi-continuous set-valued function from a metric space X to a Banach space Y taking only nonempty convex values. Let r be real and positive. Then there is a continuous selector for the expanded set-valued function*

$$B(F(\cdot); r).$$

Proof. The family of open balls

$$\{B(y) : y \in Y\},$$

with

$$B(y) = B(y; r) = \{z : \|z - y\| < r\},$$

is an open cover of Y. Since F is lower semi-continuous with nonempty values, the family of all sets of the form

$$F^{-1}(B(y)) = \{x \in X : F(x) \cap B(y) \neq \emptyset\},$$

with $y \in Y$ is an open cover of X.

By Lemma 1.1, there will be a family

$$\{p_\gamma : \gamma \in \Gamma\}$$

of continuous functions from X to $[0, 1]$ forming a partition of unity on X refining the family

$$\{F^{-1}(B(y)) : y \in Y\}.$$

Without loss of generality we may suppose that no function p_γ is identically zero.

For each γ in Γ there is a point y_γ in Y with

$$\{x : p_\gamma(x) > 0\} \subset F^{-1}(B(y_\gamma)).$$

Then, for all x with $p_\gamma(x) > 0$, we have

$$x \in F^{-1}(B(y_\gamma)),$$

so that

$$y_\gamma \in B(F(x); r).$$

Now consider the function $f : X \longrightarrow Y$ defined by

$$f(x) = \sum \{p_\gamma(x)y_\gamma : \gamma \in \Gamma\}.$$

For each ξ in X there is a neighborhood N_ξ of ξ and a finite subset Φ of Γ with

$$p_\gamma(x) = 0 \quad \text{for} \quad x \in N_\xi \quad \text{and} \quad \gamma \in \Gamma \backslash \Phi.$$

Thus

$$f(x) = \sum \{p_\varphi(x)y_\varphi : \varphi \in \Phi\}$$

for all x in N_ξ. Hence f is well defined on N_ξ and is continuous at ξ. This shows that f is well defined and continuous on X.

Since, for all x in X,

$$f(x) = \sum \left\{p_\gamma(x)y_\gamma : \gamma \in \Gamma \text{ and } p_\gamma(x) > 0\right\},$$

and

$$y_\gamma \in B(F(x); r), \quad \text{when } p_\gamma(x) > 0,$$

it follows that $f(x)$ is a finite convex combination of points in $B(F(x); r)$ and so belongs to this convex set. Thus f is the required continuous selector for $B(F(x); r)$. $\square$

Proof of Theorem 1.1. The plan is to construct a sequence of set-valued functions

$$F_0 = F, \ F_1, \ F_2, \ldots$$

and a sequence of continuous functions

$$f_1, \ f_2, \ f_3, \ldots$$

in the order

$$F_0 = F, \ f_1, \ F_1, \ f_2, \ F_2, \ldots$$

satisfying the following conditions:

(1) F_i is a lower semi-continuous set-valued function with nonempty convex values for $i \geq 0$;

(2) $F_i(x) \subset F_{i-1}(x)$, $i \geq 1$;

(3) $\operatorname{diam} F_i(x) \leq 2^{-i+1}$, $i \geq 1$;

(4) f_{i+1} is a continuous selector for $B(F_i(\cdot); 2^{-i-1})$, $i \geq 0$;

(5) $F_{i+1}(x) = F_i(x) \cap B(f_{i+1}(x); 2^{-i-1})$, for $x \in X$ and $i \geq 0$.

Once this construction is complete it will be easy to show that the functions $f_1, f_2, \ldots$ converge to the required selector for F.

Clearly the choice $F_0 = F$ satisfies (1) by our hypotheses and satisfies the conditions (2) and (3) which are vacuous when $i = 0$.

Suppose that for some $i \geq 0$, the set-function F_i has been chosen satisfying the conditions (1), (2) and (3). Using Lemma 1.2 we choose a continuous selector f_{i+1} for the expanded set-function $B(F_i(\cdot); 2^{-i-1})$, ensuring that (4) holds for this i.

We now take

$$F_{i+1}(x) = F_i(x) \cap B\!\left(f_{i+1}(x); 2^{-i-1}\right),$$

for $x \in X$, in order to satisfy (5), for this i. We have to verify that this choice satisfies (1), (2) and (3) for $i + 1$. Since

$$f_{i+1}(x) \in B\!\left(F_i(x); 2^{-i-1}\right),$$

we have

$$\|f_{i+1}(x) - y\| < 2^{-i-1}$$

for some $y \in F_i(x)$. Thus

$$F_i(x) \cap B\!\left(f_{i+1}(x); 2^{-i-1}\right) \neq \emptyset.$$

Hence F_{i+1} takes nonempty values that are convex, since $F_i(x)$ and $B(f_{i+1}(x); 2^{-i-1})$ are both convex. We also have

$$\operatorname{diam} F_{i+1}(x) \leq \operatorname{diam} B\!\left(f_{i+1}(x); 2^{-i-1}\right)$$

$$\leq 2^{-i}.$$

To obtain (1), (2) and (3) for $i + 1$ it remains to prove that F_{i+1} is lower semi-continuous. This needs a special argument.

Let G be any open set in Y. We need to show that $F_{i+1}^{-1}(G)$ is open. Consider any point ξ in $F_{i+1}^{-1}(G)$. Then there is a point η of Y with

$$\eta \in F_{i+1}(\xi) \cap G$$

$$= F_i(\xi) \cap B\big(f_{i+1}(\xi); 2^{-i-1}\big) \cap G.$$

Now η belongs to

$$F_i(\xi) \cap B\big(f_{i+1}(\xi); 2^{-i-1} - \lambda\big) \cap G$$

for some suitably chosen $\lambda > 0$. Since F_i is lower semi-continuous and

$$B\big(f_{i+1}(\xi); 2^{-i-1} - \lambda\big) \cap G$$

is an open set meeting $F_i(\xi)$, the set

$$U = F_i^{-1}\big(B(f_{i+1}(\xi); 2^{-i-1} - \lambda) \cap G\big)$$

is an open set containing ξ. Since f_{i+1} is continuous, the set

$$V = \{x : \|f_{i+1}(\xi) - f_{i+1}(x)\| < \lambda\}$$

is also an open set containing ξ. For $x \in V$, we have

$$B\big(f_{i+1}(x), 2^{-i-1}\big) \supset B\big(f_{i+1}(\xi), 2^{-i-1} - \lambda\big).$$

Thus for $x \in U \cap V$, we have

$$F_{i+1}(x) \cap G = F_i(x) \cap B\big(f_{i+1}(x); 2^{-i-1}\big) \cap G$$

$$\supset F_i(x) \cap B\big(f_{i+1}(\xi), 2^{-i-1} - \lambda\big) \cap G$$

$$\neq \emptyset$$

using the definitions of U. This shows that $F_{i+1}^{-1}(G)$ contains the open set $U \cap V$ containing ξ. Hence $F_{i+1}^{-1}(G)$ is open and F_{i+1} is lower semi-continuous as required. The inductive construction is now complete.

The conditions (4) and (2) ensure that

$$f_{i+1}(x) \in B\big(F_i(x); 2^{-i-1}\big) \subset B\big(F_{i-1}(x), 2^{-i-1}\big),$$

$$f_i(x) \in B\big(F_{i-1}(x); 2^{-i}\big),$$

so that $f_i(x)$, $f_{i+1}(x)$ both belong to the set $B(F_{i-1}(x); 2^{-i})$ which by (3) has diameter at most

$$2^{-i+2} + 2 \times 2^{-i} < 2^{-i+3}.$$

This ensures that the sequence $f_1, f_2, \ldots$ converges uniformly to a function, f say, that is necessarily continuous. The conditions (4) and (2) ensure that

$$f_i(x) \in B\left(F(x); 2^{-i}\right).$$

Since $F(x)$ is closed we necessarily have $f(x) \in F(x)$, and we have the required continuous selector. □

Proof of Theorem 1.2. Since u is a linear map of Y onto X, each set $u^{-1}(x)$, with $x \in X$, is nonempty and convex. Since u is continuous, each set $u^{-1}(x)$, with $x \in X$, is also closed. Since u is a continuous linear map Y onto X, the open mapping theorem, see for example [69], applies and $U(G)$ is open in X whenever G is open in Y. Hence, for each open G in Y,

$$\left\{x : u^{-1}(x) \cap G \neq \emptyset\right\} = \{x : x \in u(G)\}$$

is open in X. Thus $u^{-1}(x)$ is a lower semi-continuous set-valued map from X to Y taking only nonempty closed convex values. Now the conditions of Michael's selection theorem are satisfied and Theorem 1.2 follows. □

It will be useful at this stage to obtain two further consequences of Michael's theorem.

Theorem 1.3   *Let C be a closed convex set in a Banach space Y and let ϵ be positive. Then there is a retraction γ of Y to C with*

$$\|\gamma(y) - y\| \leq (1 + \epsilon) \inf\{\|c - y\| : c \in C\}$$

for all y in Y.

Theorem 1.4   *Let f be a function of the first Baire class from a metric space X to a Banach space Y and suppose that f takes its values in a closed convex set C in Y. Then f is also of the first Baire class when regarded as a map from X to C.*

Proof of Theorem 1.3. Write

$$\rho(y) = \inf_{c \in C} \|c - y\|$$

and

$$\overline{B}(y) = \{z : \|z - y\| \leq (1 + \epsilon)\rho(y)\}$$

for all y in Y. We wish to prove that

$$\overline{B}(y) \cap C$$

is a lower semi-continuous set-valued function. It clearly has nonempty closed convex values.

Consider any y_0 in Y and any open set G that meets

$$\overline{B}(y_0) \cap C.$$

Choose $\eta > 0$ and c_0 in $\overline{B}(y_0) \cap C \cap G$ so that

$$B(c_0; \eta) \subset G \; (c_0 = y_0 \text{ if } c_0 \text{ is in } C).$$

Then

$$\|c_0 - y_0\| \leq (1 + \epsilon)\rho(y_0).$$

Choose c_1 in C with

$$\|c_1 - y_0\| \leq \left(1 + \frac{1}{2}\epsilon\right)\rho(y_0),$$

($c_1 = c_0 = y_0$ if y_0 is in C). Consider first the case when $c_1 \neq c_0$. In this case we write

$$c_2 = (1 - \theta)c_0 + \theta c_1,$$

with

$$\theta = \min\left\{1, \frac{\frac{1}{2}\eta}{\|c_1 - c_0\|}\right\}.$$

Then $0 < \theta \leq 1$ and

$$\|c_2 - c_0\| = \|\theta(c_1 - c_0)\| \leq \frac{1}{2}\eta,$$

so that $c_2 \in G$. Also

$$\begin{aligned}
\|y_0 - c_2\| &= \|(1 - \theta)(y_0 - c_0) + \theta(y_0 - c_1)\| \\
&\leq (1 - \theta)\|y_0 - c_0\| + \theta\|y_0 - c_1\| \\
&\leq \left[(1 - \theta)(1 + \epsilon) + \theta\left(1 + \frac{1}{2}\epsilon\right)\right]\rho(y_0) \\
&= \left(1 + \epsilon - \frac{1}{2}\theta\epsilon\right)\rho(y_0).
\end{aligned}$$

Since

$$|\rho(y) - \rho(y_0)| \leq \|y - y_0\|,$$

for all y in Y, we have

$$\|y - c_2\| \leq (1 + \epsilon)\rho(y)$$

and

$$\overline{B}(y) \cap C \cap G \neq \emptyset,$$

for all y sufficiently close to y_0. We reach the same conclusion, by a simpler version of this argument, in the case when $c_1 = c_0$ by taking $c_2 = c_0$, and noting that $c_2 \in C \cap G$ and

$$\|c_2 - y_0\| \le \left(1 + \frac{1}{2}\epsilon\right)\rho(y_0),$$

so that

$$\|c_2 - y\| \le (1 + \epsilon)\rho(y),$$

for all y sufficiently close to y_0, but not in C. In the case that $y_0 \in C$, for $y \in C$ sufficiently close to y_0, we also get that $\overline{B}(y) \cap C \cap G \neq \emptyset$. Thus $\overline{B}(\cdot) \cap C$ is lower semi-continuous.

Using Michael's Theorem 1.1 we take $\gamma(y)$ to be a continuous selector for the set-valued map F defined by

$$F(y) = \overline{B}(y) \cap C.$$

This $\gamma(y)$ is a retraction of Y to C with the required property. $\quad\square$

Proof of Theorem 1.4. Since f is of the first Baire class, f is the pointwise limit of a sequence $f_1, f_2, \ldots$ of continuous functions from X to Y. By Theorem 1.3 we can take γ to be a retraction from Y to C. Since a retraction is continuous, each function $\gamma \circ f_i$ is continuous from X to C. Again, for each x in X,

$$\gamma(f_i(x)) \to \gamma(f(x)) = f(x)$$

as $i \to \infty$. Thus f is of the first Baire class when regarded as a function from X to C. $\quad\square$

1.2 RESULTS OF KURATOWSKI AND RYLL-NARDZEWSKI

In this section we establish an abstract selection theorem of Kuratowski and Ryll-Nardzewski and we draw from it one of the various corollaries that they obtain.

We say that a family $\mathcal{L}$ of sets in a space X is a *field* if

$$X \backslash A, \quad A \cup B \quad \text{and} \quad A \cap B$$

belong to $\mathcal{L}$ whenever A and B belong to $\mathcal{L}$. We use $\sigma\mathcal{L}$ to denote the family of all countable unions of sets from $\mathcal{L}$. For example, if X is a metric space and $\mathcal{L}$ is the family of all sets that are both $\mathcal{F}_\sigma$-sets and $\mathcal{G}_\delta$-sets, then $\mathcal{L}$ is a field and $\sigma\mathcal{L}$ is the family of all $\mathcal{F}_\sigma$-sets in X.

Theorem 1.5 (Kuratowski and Ryll-Nardzewski) *Let X be any space and* *let $\mathcal{L}$ be a field of sets of X. Let Y be a complete separable metric space. Let F be a set-valued function from X to Y, taking only nonempty closed values, and such that*

$$F^{-1}(G) \in \sigma\mathcal{L},$$

whenever G is open in Y. Then F has a selector f with the property

$$f^{-1}(G) \in \sigma\mathcal{L},$$

whenever G is open in Y.

This has an immediate consequence.

Corollary 1.1 (Kuratowski and Ryll-Nardzewski) *Let X be a metric space and let Y be a complete separable metric space. Let F be a lower semicontinuous set-valued function from X to Y, taking only nonempty closed values. Then F has a selector of the first Borel class.*

Before we prove the main theorem we need a lemma.

Lemma 1.3 *Let S be a countably additive family of sets of a space X. Let Y be a metric space. Let $f_1, f_2, \ldots$ be a sequence of functions from X to Y converging uniformly to a function f. Suppose that, for $n \geq 1$,*

$$f_n^{-1}(G) \in S,$$

whenever G is open in Y. Then

$$f^{-1}(G) \in S,$$

whenever G is open in Y.

Proof. Let d be the metric on Y. By the uniform convergence we can choose integers

$$m(n), \quad n \geq 1,$$

with $m(n) > n$, and

$$\sup\{d(f_{m(n)}(x), f(x)) : x \in X\} \leq 1/n,$$

for $n \geq 1$.

Consider any open set G in Y. Write

$$G = \bigcup_{n=1}^{\infty} G_n,$$

with

$$G_n = \{y : d(y, Y \setminus G) > 1/n\}$$

open in Y. We verify that

$$f^{-1}(G) = \bigcup_{n=1}^{\infty} f_{m(n)}^{-1}(G_n).$$

If $\xi \in f^{-1}(G)$, then

$$f(\xi) \in G,$$

so that

$$f_{m(n)}(\xi) \in G_n,$$

for all sufficiently large n, and

$$\xi \in \bigcup_{n=1}^{\infty} f_{m(n)}^{-1}(G_n).$$

Now suppose that

$$\xi \in f_{m(n)}^{-1}(G_n)$$

for some $n \geq 1$. Then we have

$$d(f_{m(n)}(\xi) , Y \setminus G) > 1/n,$$

as well as

$$d(f_{m(n)}(\xi), f(\xi)) \leq 1/n,$$

so that $f(\xi) \in G$ and $\xi \in f^{-1}(G)$. By the condition on the functions f_n, $n \geq 1$, the set

$$f^{-1}(G) = \bigcup_{n=1}^{\infty} f_{m(n)}^{-1}(G_n)$$

belongs to the countably additive family S, as required. $\square$

Proof of Theorem 1.5. By a trivial change of the metric d on Y we may suppose that Y has diameter less than 1. Since Y is separable we can choose a sequence $r_1, r_2, \ldots$ of points dense in Y. We obtain the selector f as the uniform limit of a sequence $f_0, f_1, \ldots$ of functions from X to Y, each function taking its values from the sequence $r_1, r_2, \ldots$. We construct this sequence to satisfy the conditions:

(a) $f_n^{-1}(G) \in \sigma\mathcal{L}$, whenever G is open in Y and $n \geq 0$;

(b) $d(f_n(x), F(x)) < 2^{-n}$ for all x and $n \geq 0$; and

(c) $d(f_n(x), f_{n-1}(x)) < 2^{-n+1}$ for all x and $n \geq 1$.

We start the construction by taking $f_0(x) = r_1$ for all x. Then either $r_1 \notin G$ and

$$f_0^{-1}(G) = \emptyset = F^{-1}(\emptyset),$$

or $r_1 \in G$, and

$$f^{-1}(G) = X = F^{-1}(X)$$

and $f^{-1}(G) \in \sigma(L)$ for every subset G of Y. Further

$$d(f_0(x), F(x)) < 1,$$

as Y has diameter less than 1. Thus the conditions (a) and (b) are satisfied when $n = 0$.

Now suppose that $n \geq 1$ and that f_{n-1} has been chosen to satisfy the conditions (a) and (b). Write

$$C_i^{(n)} = F^{-1}(B(r_i; 2^{-n})), \tag{1.1}$$

$$D_i^{(n)} = f_{n-1}^{-1}\left(B(r_i; 2^{-n+1})\right), \tag{1.2}$$

$$A_i^{(n)} = C_i^{(n)} \cap D_i^{(n)}. \tag{1.3}$$

Then $C_i^{(n)}$ and $D_i^{(n)}$, and so also $A_i^{(n)}$, belong to σL. We explain that we have chosen these sets to ensure that, if x belongs to $A_i^{(n)}$ and f_n is assigned the value r_i at x, then, as we shall see, the conditions (b) and (c) will be satisfied.

We verify that

$$X = \bigcup_{i=1}^{\infty} A_i^{(n)}.$$

For each ξ in X, condition (b) ensures that there is a point η of $F(\xi)$ with

$$d(\eta, f_{n-1}(\xi)) < 2^{-n+1}.$$

Since the sequence $r_1, r_2, \ldots$ is dense in Y, we can choose i with

$$d(r_i, \eta) < \min\left\{2^{-n}, 2^{-n+1} - d(\eta, f_{n-1}(\xi))\right\}.$$

This ensures that

$$d(r_i, f_{n-1}(\xi)) < 2^{-n+1}$$

and

$$d(r_i, F(\xi)) \leq d(r_i, \eta) < 2^{-n}.$$

Thus

$$\xi \in F^{-1}(B(r_i, 2^{-n})) = C_i^{(n)}, \tag{1.4}$$

$$\xi \in f_{n-1}^{-1}\left(B(r_i, 2^{-n+1})\right) = D_i^{(n)} \tag{1.5}$$

and $\xi \in A_i^{(n)}$. Hence

$$X = \bigcup_{i=1}^{\infty} A_i^{(n)},$$

as required.

At this stage it is tempting to define f_n by assigning the value r_i to f_n at x with i the smallest integer with $x \in A_i^{(n)}$. If we did this we would satisfy (b) and (c); however, we would not obtain (a). To obtain (a), (b) and (c) simultaneously Kuratowski and Ryll-Nardzewski have to be more subtle.

Since $A_i^{(n)} \in \sigma\mathcal{L}$, we can write

$$A_i^{(n)} = \bigcup_{j=1}^{\infty} E_{i,j}^{(n)}$$

with each set $E_{i,j}^{(n)}$ in $\mathcal{L}$. Let $E_k^{(n)}$, $k = 1, 2, \ldots$ be a rearrangement of the double sequence

$$E_{i,j}^{(n)}, \quad i \geq 1, \ j \geq 1,$$

as a single sequence. Write

$$L_k^{(n)} = E_k^{(n)} \setminus \bigcup_{\ell < k} E_\ell^{(n)}.$$

Then X is the disjoint union

$$X = \bigcup_{k=1}^{\infty} L_k^{(n)}$$

of sets belonging to $\mathcal{L}$.

For each k, let $E_{i(k),j(k)}^{(n)}$ be the original name for the set $E_k^{(n)}$. We define the function f_n by taking

$$f_n(x) = r_{i(k)}, \quad \text{when } x \in L_k^{(n)}.$$

Now, if G is any set in Y (open or not),

$$f_n^{-1}(G) = \bigcup\left\{L_k^{(n)} : r_{i(k)} \in G\right\},$$

which is a set of $\sigma\mathcal{L}$. Thus (a) holds for this n. Further, for ξ in X, we can choose κ with $\xi \in L_\kappa^{(n)}$. Then

$$\xi \in E_\kappa^{(n)} = E_{i(\kappa),j(\kappa)}^{(n)} \subset A_{i(\kappa)}^{(n)} = C_{i(\kappa)}^{(n)} \cap D_{i(\kappa)}^{(n)}$$

and

Math

send for LABELING.
```
=====================
```
[ ] *BEFORE* Catalog.
    [ ] CADM
    [ ] COPY CAT.
```
=+++=================
```
TREATMENT DECISION
Init'l & Date: _____
_____
_____
_____
_____
```
=====================
```
[ ] return to BIND/PRES.
[ ] treated
```
======++=============
```
Send to LABELING.
```
====++===============
```
SPECIAL NOTES
Init'l: _____

```
=====================
```
c:\msword\routing.doc
11.18.91

$$f_n(\xi) = r_{i(\kappa)},$$

so that, by (1.4) and (1.5),

$$\xi \in C^n_{i(\kappa)} = F^{-1}\big(B(r_{i(\kappa)}; 2^{-n})\big)$$

and

$$\xi \in D^n_{i(\kappa)} = f^{-1}_{n-1}\big(B(r_{i(\kappa)}; 2^{-n+1})\big).$$

Hence

$$d\big(f_n(\xi), F(\xi)\big) = d\big(r_{i(\kappa)}, F(\xi)\big) < 2^{-n}$$

and

$$d\big(f_n(\xi), f_{n-1}(\xi)\big) = d\big(r_{i(\kappa)}, f_{n-1}(\xi)\big) < 2^{-n+1}.$$

Thus the conditions (b) and (c) also hold. This completes the construction.

By the condition (c) and the completeness of Y, the sequence of functions $f_0, f_1, \ldots$ converges uniformly to a function, say f, from X to Y. Since $F(x)$ is closed for each x in X, the condition (b) ensures that f is a selector for F. By Lemma 1.3, with $S = \sigma\mathcal{L}$, f satisfies the requirement that $f^{-1}(G) \in \sigma\mathcal{L}$ whenever G is open in Y. $\square$

1.3 REMARKS

1. Michael proves his selection theorem in the more general case when X is a paracompact space. The version of Michael's proof given can easily be modified to cover this case; it is enough to note that Lemma 1.1 holds for any paracompact space X.

2. E. Michael [54] also obtains the following remarkable converse to this theorem.
   Let X be a T_1 topological space. If every lower semi-continuous set-valued function F from X to a Banach space Y, taking only nonempty closed convex values, has a continuous selector, then X is necessarily paracompact.

3. The following simple example shows that the convexity of the values of the set-valued function is essential for the validity of Michael's theorem. Take F to be the set-valued function from the open interval $(0, 1)$ to the real line $\mathbb{R}$ defined by

$$F(x) = \{0\}, \quad \text{if } 0 < x \le \frac{1}{3}, \tag{1.6}$$

$$F(x) = \{0, 1\}, \quad \text{if } \frac{1}{3} < x < \frac{2}{3}, \tag{1.7}$$

$$F(x) = \{1\}, \quad \text{if } \frac{2}{3} \le x < 1. \tag{1.8}$$

It is easy to verify that F is a lower semi-continuous set-valued function with nonempty compact values that has no continuous selector.

4. Lower semi-continuous set functions have a peculiar property. Let F be a lower semi-continuous set-valued function from X to Y. Let $H(x)$ be a dense subset of $F(x)$ chosen separately for each x in X. No matter how H is chosen in this way, it is lower semi-continuous. For this reason one usually works with lower semi-continuous set-valued functions that only take closed values, or that are "fat" like the function $B(F(\cdot), r)$ considered in Lemma 1.2.

5. Let Y be a Banach space taken with its norm metric. Let X be the space of nonempty closed bounded convex sets in Y with the Hausdorff metric. Let I be the set-valued function from X to Y that assigns to each nonempty closed bounded convex set K in Y, regarded as a point of X, the set

$$I(K) = K$$

regarded as a set in Y. Then I is a lower semi-continuous set-valued function taking only nonempty closed (bounded) convex values. Hence there is a continuous selector $k : X \to Y$ for I that assigns to each nonempty closed bounded convex sets in Y one of its points.

6. The problem of the continuous selection of a point from a nonempty bounded closed convex set in $\mathbb{R}^d$ is of considerable geometric interest. We use the notation of Remark 5.

(a) The most "obvious" selector is perhaps to take $k(K)$ to be the center of gravity or centroid of the nonempty convex set K in $\mathbb{R}^d$. However, this selector is not continuous even in the case $d = 2$. Let $T(\theta)$ be the, sometimes degenerate, triangle defined by

$$0 \le x \le 1, \qquad -x \tan \theta \le y \le x \tan \theta,$$

for $0 \le \theta \le \frac{1}{4} \pi$. When $\theta = 0$, the center of gravity is the point $(\frac{1}{2}, 0)$, when $0 < \theta \le \frac{1}{4} \pi$, the center of gravity is the point $(\frac{2}{3}, 0)$.

(b) A better choice in $\mathbb{R}^d$ is to take $k(K)$ to be the center of the sphere of minimal radius that contains K. The sphere of minimal radius is unique and $k(K)$ is a point of K that varies continuously with K. This selector has the disadvantage that even when K has an inner point, $k(K)$ may lie on the boundary of K.

(c) A bad choice in $\mathbb{R}^d$ is to take $k(K)$ to be the center of the maximal spherical ball (of some dimension) contained in K. This point is not in general well defined (consider a rectangle in $\mathbb{R}^2$).

(d) An elegant choice is the Steiner point. Let K be a nonempty closed bounded convex point in $\mathbb{R}^d$. Let

$$H(\mathbf{u}) = \sup\{\langle \mathbf{x}, \mathbf{u} \rangle : \mathbf{x} \in K\}$$

be the support function of K. Take

$$\mathbf{s}(K) = \frac{\int_{S^{d-1}} \mathbf{u} H(\mathbf{u})\, dw}{\int_{S^{d-1}} \langle \mathbf{a}, \mathbf{u} \rangle\, dw},$$

with dw the $(d-1)$-dimensional volume element on the unit sphere S^{d-1} in $\mathbb{R}^d$ and with $\mathbf{a}$ a fixed unit vector, its direction being irrelevant. Then $\mathbf{s}(K)$ is the Steiner point of K. An alternative definition that gives a clearer idea of the construction of $\mathbf{s}$ that may be used when K is strictly convex. For each unit vector $\mathbf{u}$ let $\mathbf{c}(\mathbf{u})$ be the point of contact of the hyperplane perpendicular to $\mathbf{u}$ touching K at a point $\mathbf{c}(\mathbf{u})$ admitting $\mathbf{u}$ as an outward normal to K. Then

$$\mathbf{s}(K) = \frac{\int_{S^{d-1}} \mathbf{c}(\mathbf{u})\, dw}{\int_{S^{d-1}}\, dw}.$$

The Steiner point varies continuously with K and always lies in the relative interior of K, when K has positive dimension. See, for example, [11] or [16, chapter 14].

(e) Although the choices (b) and (d) move with K under any rigid motion, they are not invariant under affine transformations. Indeed no continuous selector can be invariant under affine transformations. Such a selector would have to assign the center of gravity to any equilateral triangle, and so also to any triangle. It would also have to assign the midpoint to any line segment. This is impossible by (a).

(f) For a more sophisticated approach to such selections in Banach spaces see chapter 4.

7. Michael [54] states the Bartle–Graves Theorem explicitly, attributing it to Bartle and Graves [3]. However, although their paper contains many interesting results, it is not easy to find this theorem in their paper.

8. If f is a function of Baire class α from a metric space X to a Banach space Y, taking only values in a closed convex set Y, then f is also of Baire class α when regarded as a map from X to C. This may be proved by a modification of the proof of Theorem 1.4.

9. In the Corollary to Theorem 1.5, it is not essential for X to be a metric space, it is enough to suppose that X is a Hausdorff space where each closed set is a G_δ-set.

10. We quote another of the consequences that Kuratowski and Ryll-Nard-zewski draw from their theorem. Let α be a countable ordinal. Let F be a set-valued function from a metric space X to a complete separable metric space Y, taking only nonempty closed values, with the property that $F^{-1}(G)$ is of additive Borel class α for each open set G in Y. Then F has a selector f of Borel class α. This is obtained easily from Theorem 1.5 by taking $\mathcal{L}$ to be the family of sets in X of ambiguous class α. Kuratowski and Ryll-Nardzewski obtain further interesting corollaries in their note. Hansell [20] gives a number of results extending the work of Kuratowski and Ryll-Nardzewski to nonseparable spaces. See also the result of Srivatsa, Theorem 6.4 below.

11. We now analyze the version of Michael's proof of his continuous selection theorem from the point of view expressed in the introduction. Our starting point is a metric space X, a Banach space Y and a lower semi-continuous set-valued function F that takes only nonempty closed convex values in Y. During the course of the proof we find that we need to study new set-valued functions constructed inductively from F, with the same properties as F. The proof becomes purely constructive once we have proved Lemma 1.2, not only for the given set-valued function F, but also for the set-valued functions constructed inductively from F. In Lemma 1.2, the starting point is the metric space X, the Banach space Y, the lower semi-continuous set-valued function F from X to Y taking only nonempty closed convex values in Y, and the positive real number r. In the proof, one introduces the family

$$\{B(y) : y \in Y\}$$

of open balls in Y. By the lower semi-continuity of F the family

$$\{F^{-1}(B(y)) : y \in Y\}$$

is an open cover of X. Using Lemma 1.1 there will be a partition of unity

$$\{p_\gamma : \gamma \in \Gamma\}$$

on X refining the family

$$\{F^{-1}(B(y)) : y \in Y\}$$

so that for each γ in Γ, there is a y_γ in Y, with

$$\{x : p_{\gamma(x)} > 0\} \subset F^{-1}\big(B(y_\gamma)\big).$$

However, to obtain this partition of unity requires extensive use of the axiom of choice. Once the extra structure has been found in this way, the rest of the proof of the lemma is purely constructive. As we have remarked, the deduction of the theorem from the lemma is also constructive.

Chapter 2

Functions that are constant on the sets of a disjoint discretely σ-decomposable family of $\mathcal{F}_\sigma$-sets

Most of the selectors that we construct in subsequent chapters are obtained as uniform limits of certain "approximate selectors" that are point-valued functions which are constant on the sets of some partition of a space into a disjoint discretely σ-decomposable family of $\mathcal{F}_\sigma$-sets. Our aim in this chapter is to define and construct such partitions of space and to obtain properties of such functions and their limits. The main conclusions are summarized in Theorem 2.1.

First we need to define one term and to derive some results in the theory of nonseparable metric spaces as developed by Montgomery [59], Stone [76–78] and Hansell [19].

A family $\{S_\gamma : \gamma \in \Gamma\}$ of sets in a topological space X is said to be *discrete* if each point x of X has a neighborhood N_x such that

$$N_x \cap S_\gamma \neq \emptyset,$$

for at most one γ in Γ. A family $\{S_\gamma : \gamma \in \Gamma\}$ is said to be *discretely σ-decomposable* if it is possible to write

$$S_\gamma = \bigcup_{n=1}^{\infty} S_\gamma^{(n)}, \quad \text{for } \gamma \in \Gamma,$$

with each family

$$\{S_\gamma^{(n)} : \gamma \in \Gamma\}, \quad n \geq 1,$$

discrete in X.

2.1 DISCRETELY σ-DECOMPOSABLE PARTITIONS OF A METRIC SPACE

In this section we develop results concerning discretely σ-decomposable partitions of a metric space into $\mathcal{F}_\sigma$-sets. We prove a series of lemmas.

Lemma 2.1 *Let Γ be an ordinal and let*

$$\{G_\gamma : \gamma \in \Gamma\}$$

be an open cover of a metric space X. Write

$$U_\gamma = G_\gamma \backslash \bigcup \{G_\beta : 0 \le \beta < \gamma\},$$

for $\gamma \in \Gamma$. Then

$$\{U_\gamma : \gamma \in \Gamma\}$$

is a disjoint discretely σ-decomposable cover of X by $\mathcal{F}_\sigma$-sets.

Proof. Let d be the metric on X. For each γ in Γ and each $n \ge 1$, write

$$G_\gamma^{(n)} = \{x : B(x; 2^{-n}) \subset G_\gamma\},$$

where we use $B(x; r)$ to denote

$$B(x; r) = \{y : d(y, x) < r\}.$$

Then $G_\gamma^{(n)}$ is closed for $n \ge 1$, and

$$G_\gamma = \bigcup_{n=1}^{\infty} G_\gamma^{(n)}.$$

Write

$$U_\gamma^{(n)} = G_\gamma^{(n)} \backslash \bigcup \{G_\beta : 0 \le \beta < \gamma\}$$

and

$$U_\gamma = \bigcup_{n=1}^{\infty} U_\gamma^{(n)}.$$

Then each set $U_\gamma^{(n)}$ is closed and

$$\{U_\gamma : \gamma \in \Gamma\}$$

is a disjoint cover of X by $\mathcal{F}_\sigma$-sets.

Suppose $n \ge 1$ and consider the family

$$\{U_\gamma^{(n)} : \gamma \in \Gamma\}.$$

If $x \in X$, it may happen that the neighborhood $B(x; 2^{-n-1})$ of x meets none of the sets

$$U_\gamma^{(n)}, \quad \gamma \in \Gamma.$$

If this does not happen, there will be a smallest ordinal γ^* for which

$$B(x; 2^{-n-1}) \cap U_\gamma^{(n)} \ne \emptyset.$$

In this case there is a point y in $U_{\gamma^*}^{(n)}$ with

$$d(x, y) < 2^{-n-1}.$$

Now $y \in G_{\gamma^*}^{(n)}$ and

$$B(y; 2^{-n}) \subset G_{\gamma^*}.$$

Hence

$$B(x; 2^{-n-1}) \subset G_{\gamma^*},$$

ensuring that

$$B(x; 2^{-n-1}) \cap U_\gamma^{(n)} = \emptyset,$$

for $\gamma^* < \gamma < \Gamma$, as well as for $0 \le \gamma < \gamma^*$. This shows that the family

$$\{U_\gamma^{(n)} : \gamma \in \Gamma\}$$

is discrete in X, as required. $\square$

Lemma 2.2 *Let $\{U_\gamma : \gamma \in \Gamma\}$ be a disjoint discretely σ-decomposable family of $\mathcal{F}_\sigma$-sets in a topological space X. Then, for each γ in Γ, we can write*

$$U_\gamma = \bigcup_{i=1}^{\infty} F_\gamma^{(i)}, \quad \gamma \in \Gamma,$$

with

$$F_\gamma^{(1)} \subset F_\gamma^{(2)} \subset \cdots, \quad \gamma \in \Gamma,$$

and with the family

$$\{F_\gamma^{(i)} : \gamma \in \Gamma\}$$

a discrete family of closed sets for $i \ge 1$.

Proof. Since the family $\{U_\gamma : \gamma \in \Gamma\}$ is a family of $\mathcal{F}_\sigma$-sets, we can write

$$U_\gamma = \bigcup_{j=1}^{\infty} H_\gamma^{(j)}, \quad \gamma \in \Gamma,$$

with each set $H_\gamma^{(j)}, j \ge 1, \gamma \in \Gamma$, closed in X. Since the family $\{U_\gamma : \gamma \in \Gamma\}$ is discretely σ-decomposable we can write

$$U_\gamma = \bigcup_{k=1}^{\infty} I_\gamma^{(k)}, \quad \gamma \in \Gamma,$$

with each family

$$\{I_\gamma^{(k)} : \gamma \in \Gamma\}, \quad k \geq 1$$

discrete in X. Then each family

$$\{\mathrm{cl}\, I_\gamma^{(k)} : \gamma \in \Gamma\}, \quad k \geq 1$$

is also discrete in X. Write

$$J_\gamma^{(j,k)} = H_\gamma^{(j)} \cap \mathrm{cl}\, I_\gamma^{(k)}, \quad j,k \geq 1, \ \gamma \in \Gamma.$$

Then

$$U_\gamma = \bigcup_{j,k \geq 1} J_\gamma^{(j,k)}, \quad \gamma \in \Gamma,$$

with each family

$$\{J_\gamma^{(j,k)} : \gamma \in \Gamma\}, \quad j,k \geq 1,$$

a discrete family of closed sets.

For $i \geq 1$, write

$$F_\gamma^{(i)} = \bigcup_{j+k \leq i+1} J_\gamma^{(j,k)}, \quad \gamma \in \Gamma.$$

Then

$$F_\gamma^{(1)} \subset F_\gamma^{(2)} \subset \cdots,$$

$$\bigcup_{i=1}^{\infty} F_\gamma^{(i)} = U_\gamma$$

and each set $F_\gamma^{(i)}$, $i \geq 1$, $\gamma \in \Gamma$, is closed.

We verify that each family

$$\{F_\gamma^{(i)} : \gamma \in \Gamma\}, \quad i \geq 1,$$

is discrete in X. Consider a fixed $i \geq 1$ and a fixed x in X. First consider the case when x belongs to the set U_{γ^*} with $\gamma^* \in \Gamma$. Then x belongs to none of the sets U_γ with $\gamma \neq \gamma^*$ and so, for each pair j, k with $j + k \leq i + 1$, the point x belongs to no set in the discrete family

$$\{J_\gamma^{(j,k)} : \gamma \in \Gamma, \ \gamma \neq \gamma^*\}$$

and we can choose a neighborhood $N^{(j,k)}$ of x that meets none of these sets. Thus

$$N = \bigcap_{j+k \leq i+1} N^{(j,k)}$$

is a neighborhood of x that meets none of the sets

$$F_\gamma^{(i)} = \bigcup_{j+k \leq i+1} J_\gamma^{(j,k)}, \quad \gamma \in \Gamma, \ \gamma \neq \gamma^*.$$

Thus N meets at most one of the sets $\{F_\gamma^{(i)} : \gamma \in \Gamma\}$. Similarly, if x belongs to no set U_γ, $\gamma \in \Gamma$, then N can be chosen to meet no set of the family $\{F_\gamma^{(i)} : \gamma \in \Gamma\}$. Thus this family is discrete, as required. $\square$

Lemma 2.3 *Let X be a topological space. Let*

$$\{U_\gamma : \gamma \in \Gamma\}$$

be a disjoint discretely σ-decomposable family of $\mathcal{F}_\sigma$-sets covering X. Suppose that, for each γ in Γ,

$$\{V_{\gamma\theta} : \theta \in \Theta(\gamma)\}$$

is a disjoint discretely σ-decomposable family of $\mathcal{F}_\sigma$-sets relative to U_γ and covering U_γ. Then

$$\{V_{\gamma\theta} : \theta \in \Theta(\gamma), \ \gamma \in \Gamma\}$$

is a disjoint discretely σ-decomposable family of $\mathcal{F}_\sigma$-sets relative to X and covering X.

Proof. Since a relative $\mathcal{F}_\sigma$-set of an $\mathcal{F}_\sigma$-set is an $\mathcal{F}_\sigma$-set, it is clear that

$$\{V_{\gamma\theta} : \theta \in \Theta(\gamma), \ \gamma \in \Gamma\}$$

is a disjoint family of $\mathcal{F}_\sigma$-sets covering X. We need to prove this family is discretely σ-decomposable.

Since the family

$$\{U_\gamma : \gamma \in \Gamma\}$$

is a discretely σ-decomposable family of $\mathcal{F}_\sigma$-sets, Lemma 2.2 ensures that we can write

$$U_\gamma = \bigcup_{j=1}^{\infty} U_\gamma^{(j)}, \quad \gamma \in \Gamma,$$

with each family

$$\{U_\gamma^{(j)} : \gamma \in \Gamma\}, \quad j \geq 1,$$

a discrete family of closed sets. Similarly, for each $\gamma \in \Gamma$, we can write

$$V_{\gamma\theta} = \bigcup_{k=1}^{\infty} V_{\gamma\theta}^{(k)}, \quad \theta \in \Theta(\gamma),$$

with each family

$$\{V_{\gamma\theta}^{(k)} : \theta \in \Theta(\gamma)\}, \quad k \geq 1,$$

a relatively discrete family of relatively closed sets in U_γ. Now, for $j, k \geq 1$, the family

$$\{U_\gamma^{(j)} \cap V_{\gamma\theta}^{(k)} : \theta \in \Theta(\gamma)\}$$

is a discrete family of closed sets in X. Indeed, for $j, k \geq 1$, the whole family

$$\{U_\gamma^{(j)} \cap V_{\gamma\theta}^{(k)} : \theta \in \Theta(\gamma), \ \gamma \in \Gamma\}$$

is a discrete family of closed sets in X. To see this, consider any x in X. Choose a neighborhood N_1 of X that meets no set or just the set with $\gamma = \gamma^*$, in the family

$$\{U_\gamma^{(j)} : \gamma \in \Gamma\}.$$

If necessary, choose a second neighborhood $N_2 \subset N_1$ of x that meets no set or just the set with $\theta = \theta^*$ in the family

$$\{U_{\gamma^*}^{(j)} \cap V_{\gamma^*\theta}^{(k)} : \theta \in \Theta(\gamma^*)\}.$$

Then N_2 can only meet the one set

$$U_{\gamma^*}^{(j)} \cap V_{\gamma^*\theta^*}^{(k)}$$

in the family

$$\{U_\gamma^{(j)} \cap V_{\gamma\theta}^{(k)} : \theta \in \Theta(\gamma), \ \gamma \in \Gamma\}.$$

Hence this family is discrete in X.

Since

$$\bigcup_{j,k=1}^{\infty} U_\gamma^{(j)} \cap V_{\gamma\theta}^{(k)} = V_{\gamma\theta}, \quad \theta \in \Theta, \ \gamma \in \Gamma,$$

it is now clear that the family

$$\{V_{\gamma\theta} : \theta \in \Theta(\gamma), \ \gamma \in \Gamma\}$$

is discretely σ-decomposable. $\quad\square$

Lemma 2.4 *Let X be a topological space and suppose that*

$$\{U_\gamma : \gamma \in \Gamma\}$$

and

$$\{V_\theta : \theta \in \Theta\}$$

are both disjoint discretely σ-decomposable families of $\mathcal{F}_\sigma$-sets covering X.

Then

$$\{U_\gamma \cap V_\theta : \gamma \in \Gamma,\ \theta \in \Theta\}$$

is a disjoint discretely σ-decomposable family of $\mathcal{F}_\sigma$-sets covering X.

Proof. The result follows either by use of the method used to prove Lemma 2.3 or by applying Lemma 2.3 with

$$\Theta(\gamma) = \Theta, \quad \text{for } \gamma \in \Gamma,$$

and

$$V_{\gamma\theta} = U_\gamma \cap V_\theta, \quad \theta \in \Theta(\gamma),\ \gamma \in \Gamma. \qquad \square$$

2.2 FUNCTIONS OF THE FIRST BOREL AND BAIRE CLASSES

In this section we develop results concerning functions of the first Borel and Baire classes and also functions that are constant on the sets of a discretely σ-decomposable partition of a metric space into $\mathcal{F}_\sigma$-sets.

We start with a remark. By Lemma 1.3 with S the family of $\mathcal{F}_\sigma$-sets, we see that

$$\text{if } f_1, f_2, \ldots \text{ is a sequence}$$

of functions of the first Borel class, from a topological space X to a metric space Y, that converges uniformly to a function f, then f is of the first Borel class.

Following Hansell [19] we say that a family of sets $\mathcal{B}$ is a *base for a family of sets* $\mathcal{U}$ if each set U of $\mathcal{U}$ is the union of the sets of $\mathcal{B}$ that it contains. A family of sets $\mathcal{B}$ is said to be a *base for a function f* from a space X to a topological space Y if $\mathcal{B}$ is a base for the family

$$\{f^{-1}(G) : G \text{ open in } Y\}.$$

We say that *a function f from one metric space X to a second metric space Y is σ-discrete* if f has a σ-discrete base.

Lemma 2.5 *Let X be a topological space and let Y be a metric space. The uniform limit of a sequence of σ-discrete functions from X to Y will be σ-discrete.*

Proof. Let $\{f_n\}$ be a sequence of σ-discrete functions from X to Y converging uniformly to a function f. For each $n \geq 1$, let $\mathcal{B}_n$ be a σ-discrete base for the family

$$\{f_n^{-1}(G) : G \text{ open in } Y\}.$$

Write

$$B = \bigcup \{ \mathcal{B}_n : n \geq 1 \}.$$

We verify that $\mathcal{B}$ is a σ-discrete base for the family

$$\{ f^{-1}(G) : G \text{ open in } Y \}.$$

The family $\mathcal{B}$ is clearly σ-discrete. It suffices to prove that if G is open in Y then $f^{-1}(G)$ is the union of those sets of $\mathcal{B}$ that it contains. Consider any point x^* of $f^{-1}(G)$. Then $f(x^*) \in G$. Let ρ be the metric on Y. Choose $\epsilon > 0$ so that the ball

$$B(f(x^*); \epsilon) = \{ y : \rho(y, f(x^*)) < \epsilon \}$$

is contained in G. Choose n so large that

$$\rho(f_n(x), f(x)) < \frac{1}{3} \epsilon$$

for all x in X. Then

$$B\left(f_n(x^*); \frac{1}{3} \epsilon \right)$$

is an open set in Y containing $f_n(x^*)$. So we can choose a set B in $\mathcal{B}_n$ with $x^* \in B$ and $f_n(B) \subset B(f_n(x^*); \frac{1}{3} \epsilon)$. Now all points b in B satisfy

$$\rho(f(b), f(x^*)) \leq \rho(f(b), f_n(b)) + \rho(f_n(b), f_n(x^*)) + \rho(f_n(x^*), f(x^*))$$

$$\leq \epsilon,$$

so that

$$x^* \in B \text{ and } f(B) \subset G.$$

Thus x^* belongs to the set B of $\mathcal{B}$ contained in $f^{-1}(G)$, as required. $\square$

Lemma 2.6 *Let f be a σ-discrete function of the first Borel class from one metric space X to a second metric space Y. Then f has a σ-discrete closed base.*

Proof. Since f is σ-discrete we can choose a base for f of the form

$$\mathcal{B} = \{ B_\theta : \theta \in \Theta \},$$

with

$$\Theta = \bigcup_{n=1}^{\infty} \Theta(n),$$

and each family

$$\mathcal{B}_n = \{B_\theta : \theta \in \Theta(n)\}$$

discrete in X.

Since Y is a metric space we can choose an open σ-discrete base for the open sets of Y, say

$$\mathcal{G} = \{G_\varphi : \varphi \in \Phi\},$$

with

$$\Phi = \bigcup_{m=1}^{\infty} \Phi(m),$$

and each family

$$\mathcal{G}_m = \{G_\varphi : \varphi \in \Phi(m)\},$$

discrete in Y (disjoint would suffice).

Since f is of the first Borel class each set

$$f^{-1}(G_\varphi), \quad \varphi \in \Phi,$$

is an $\mathcal{F}_\sigma$-set, say

$$F_\varphi = \bigcup_{\ell=1}^{\infty} F_\varphi^{(\ell)}, \quad \varphi \in \Phi,$$

with each $F_\varphi^{(\ell)}$ closed in X.

We study the family of sets

$$H_{\theta\varphi}^{(\ell)} = \bar{B}_\theta \cap F_\varphi^{(\ell)}, \quad \theta \in \Theta, \ \varphi \in \Phi, \ B_\theta \subset F_\varphi, \ \ell \in \mathbb{N}.$$

Note that if φ is fixed

$$B_\theta \subset \bigcup_{\ell=1}^{\infty} \bar{B}_\theta \cap F_\varphi^{(\ell)} \subset F_\varphi,$$

whenever $B_\theta \subset F_\varphi$. Since $\mathcal{B}$ is a base for f, for each φ in Φ we have

$$f^{-1}(G_\varphi) = F_\varphi$$

$$= \bigcup \{B_\theta : \theta \in \Theta, B_\theta \subset F_\varphi\}$$

$$\subset \bigcup \{H_{\theta\varphi}^{(\ell)} : \theta \in \Theta, B_\theta \subset F_\varphi, \ \ell \in \mathbb{N}\} \subset F_\varphi.$$

Now each open set in Y is the union of the sets G_φ that it contains and the family

$$\mathcal{H} = \{H_{\theta,\varphi}^{(\ell)} : \theta \in \Theta, \ \varphi \in \Phi, \ B_\theta \subset F_\varphi, \ \ell \in \mathbb{N}\}$$

is a closed base for f. The family $\mathcal{H}$ is the countable union of the families

$$\mathcal{H}_{n,m}^{(\ell)} = \left\{ H_{\theta,\varphi}^{(\ell)} : \theta \in \Theta(n),\ \varphi \in \Phi(m),\ B_\theta \subset F_\varphi \right\},$$

$\ell, m, n \in \mathbb{N}$. We verify that each of these families is discrete in X. Note that for fixed m, the sets

$$G_\varphi : \varphi \in \Phi(m)$$

are disjoint. Hence, so are the sets

$$F_\varphi : \varphi \in \Phi(m)$$

and, for fixed ℓ, the sets

$$F_\varphi^{(\ell)} : \varphi \in \Phi(m).$$

Now, for fixed θ in $\Theta(n)$, $B_\theta \neq \emptyset$, the condition $B_\theta \subset F_\varphi$ is satisfied for at most one φ in $\Phi(m)$. Since the family

$$\{B_\theta : \theta \in \Theta(n)\}$$

is discrete, so also is the family

$$\{\bar{B}_\theta : \theta \in \Theta(n)\},$$

and so also the family

$$\left\{ \bar{B}_\theta \cap F_\varphi^{(\ell)} : \theta \in \Theta(n),\ \varphi \in \Phi(m),\ B_\theta \subset F_\varphi \right\}.$$

Thus $\mathcal{H}$ is the required σ-discrete closed base for f. $\square$

Lemma 2.7 *Let f be a function, from a metric space X to a metric space Y, having a closed σ-discrete base. Let ϵ be positive. Then there is a function g from X to Y that is constant on the sets of a discretely σ-decomposable partition of X into $\mathcal{F}_\sigma$-sets and satisfies*

$$|g(x) - f(x)| < \epsilon$$

for all x in X.

Proof. Let

$$\mathcal{B} = \{B_\theta : \theta \in \Theta\},$$

with

$$\Theta = \bigcup_{n=1}^{\infty} \Theta(n),$$

and with each family

$$\mathcal{B}_n = \{B_\theta : \theta \subset \Theta(n)\}$$

a discrete family of closed sets, be a closed σ-discrete base for f. Since $\mathcal{B}$ is a base for the family

$$\{f^{-1}(G) : G \text{ open in } Y \text{ and diam } G < \epsilon\},$$

we may remove from $\mathcal{B}$ all sets B_θ with diam $f(B_\theta) \geq \epsilon$, and we still have a base for f. We suppose that this removal process has been completed without change of notation.

For $n \geq 1$, write

$$C_n = \bigcup \{B_\theta : \theta \in \Theta(n)\}.$$

Then C_n is closed for each $n \geq 1$. Write

$$D_\theta = B_\theta \backslash \bigcup \{C_m : m < n\}$$

for all θ in $\Theta(n)$, $n \geq 1$. Then the family

$$\mathcal{D} = \{D_\theta : \theta \in \Theta\}$$

is a disjoint family of $\mathcal{F}_\sigma$-sets covering X. Further each family

$$\mathcal{D}_n = \{D_\theta : \theta \in \Theta(n)\}$$

is discrete in X. Thus $\mathcal{D}$ is a σ-discrete partition of X into $\mathcal{F}_\sigma$-sets.

For each θ in Θ, for which $D_\theta \neq \emptyset$, we have

$$\text{diam } f(D_\theta) \leq \text{ diam } f(B_\theta) < \epsilon.$$

For each such θ, choose y_θ in $f(D_\theta)$. Consider the function g defined on X by taking

$$g(x) = y_\theta, \quad \text{if } x \in D_\theta, \ \theta \in \Theta.$$

Clearly g is well defined and is constant on the sets of the σ-discrete partition of X into $\mathcal{F}_\sigma$-sets. Further, for each x in X, there is a unique θ in Θ with $x \in D_\theta$ and so $f(x)$ and $g(x) = y_\theta$ both lie in $f(D_\theta)$ and

$$|g(x) - f(x)| < \epsilon. \quad \square$$

Lemma 2.8 *Let f be a function, from a metric space X to a metric space Y, having a closed σ-discrete base. Then f is the uniform limit of a sequence of functions, from X to Y, each being constant on the sets of a corresponding discretely σ-decomposable partition of X into $\mathcal{F}_\sigma$-sets.*

Proof. For each $n \geq 1$, apply Lemma 2.7, with $\epsilon = 1/n$, to construct a function g_n, from X to Y, constant on the sets of a discretely σ-decomposable partition

$\mathcal{D}_n$ of X into $\mathcal{F}_\sigma$-sets and satisfying

$$|g_n(x) - f(x)| < 1/n$$

for all x in X. This sequence $g_1, g_2, \ldots$ converges uniformly to f and satisfies our requirements. □

Lemma 2.9 *Let X and Y be metric spaces and let f be a function from X to Y that is constant on each set of a discretely σ-decomposable partition of X into $\mathcal{F}_\sigma$-sets. Then f is a σ-discrete function of the first Borel class.*

Proof. Let f be constant on the sets of the discretely σ-decomposable partition

$$\mathcal{U} = \{U_\gamma : \gamma \in \Gamma\}$$

of X into $\mathcal{F}_\sigma$-sets. We can write

$$U_\gamma = \bigcup_{i=1}^{\infty} F_\gamma^{(i)}, \quad \gamma \in \Gamma,$$

with each set $F_\gamma^{(i)}$ closed in X and each family

$$\{F_\gamma^{(i)} : \gamma \in \Gamma\}, \quad i \geq 1,$$

discrete in X. Suppose that f is a function from X to Y that is constant on each set of $\mathcal{U}$. The family

$$\{F_\gamma^{(i)} : \gamma \in \Gamma, \ i \geq 1\}$$

is clearly a closed σ-discrete base for the function f, and furthermore for each set G in Y, open or not, $f^{-1}(G)$ is a union of sets from this family and so is an $\mathcal{F}_\sigma$-set. Thus f is a σ-discrete function of the first Borel class. □

Lemma 2.10 *Let X and Y be metric spaces. Let f be a function from X to Y that is the uniform limit of a sequence f_n, $n \geq 1$, of functions from X to Y, the function f_n being constant on each set of a discretely σ-decomposable partition $\mathcal{U}_n$ of X into $\mathcal{F}_\sigma$-sets. Then f is a σ-discrete function of the first Borel class.*

Proof. By Lemma 2.9, each function f_n is a σ-discrete function of the first Borel class. Hence the limit function is of the first Borel class. By Lemma 2.5, the limit function is σ-discrete. □

Lemma 2.11 *Let f be a function from one metric space X to another Y. If f has a σ-discrete base of closed sets, then the set of points of discontinuity of f is an $\mathcal{F}_\sigma$-set of the first Baire category in X.*

Proof. Let

$$B = \{B_\theta : \theta \in \Theta\}$$

be a closed σ-discrete base for f with

$$\Theta = \bigcup_{n=1}^\infty \Theta(n),$$

and, with each family

$$\mathcal{B}_n = \{B_\theta : \theta \in \Theta(n)\},$$

a discrete family of closed sets.

Since Y is a metric space, we can choose a family

$$\mathcal{U} = \{U_\varphi : \varphi \in \Phi\},$$

with

$$\Phi = \bigcup_{m=1}^\infty \Phi(m),$$

and with

$$\mathcal{U}_m = \{U_\varphi : \varphi \in \Phi(m)\},$$

a discrete family of open sets; this family $\mathcal{U}$ forms an open σ-discrete base for the open sets of Y.

Now f is continuous at a point x of X, if, and only if

$$x \in \text{int } f^{-1}(U),$$

whenever

$$x \in f^{-1}(U)$$

and $U \in \mathcal{U}$. Hence the set D of points of discontinuity of f takes the form

$$D = \bigcup \{f^{-1}(U_\varphi) \backslash \text{int } f^{-1}(U_\varphi) : \varphi \in \Phi\}.$$

Now, for φ in Φ, the set $f^{-1}(U_\varphi)$ takes the form

$$\bigcup \{B_\theta : \theta \in \Theta(\varphi)\},$$

where $\Theta(\varphi)$ is the set of θ in Θ for which

$$B_\theta \subset f^{-1}(U_\varphi).$$

Thus D is contained in the set

$$\bigcup \{B_\theta \backslash \text{int } B_\theta : \theta \in \Theta\} = \bigcup_{h=1}^\infty \{B_\theta \backslash \text{int } B_\theta : \theta \in \Theta(h)\}.$$

Now, for $h \geq 1$,

$$\{B_\theta \setminus \text{int } B_\theta : \theta \in \Theta(h)\}$$

is a discrete family of nowhere dense sets, and so the union of this family is nowhere dense and D is of the first Baire category in X.

To prove that D is an $\mathcal{F}_\sigma$-set, we rewrite the formula for D in the form

$$D = \bigcup_{m=1}^{\infty} \bigcup \left\{ f^{-1}(U_\varphi) \setminus \text{int } f^{-1}(U_\varphi) : \varphi \in \Phi(m) \right\}$$

$$= \bigcup_{m=1}^{\infty} \bigcup \left\{ B_\theta \setminus \text{int } f^{-1}(U_\varphi) : \varphi \in \Phi(m) \text{ and } \theta \in \Theta(\varphi) \right\}$$

$$= \bigcup_{m=1}^{\infty} \bigcup \left\{ B_\theta \setminus \text{int } f^{-1}(U_\varphi) : \varphi \in \Phi(m), \ \theta \in \Theta \text{ and } B_\theta \subset f^{-1}(U_\varphi) \right\}.$$

For fixed m, the sets $f^{-1}(U_\varphi)$, $\varphi \in \Phi(m)$ are disjoint, so that, for each θ in Θ, there is at most one φ in $\Phi(m)$ for which $B_\theta \subset f^{-1}(U_\varphi)$. Let $\Theta^{(m)}$ be the set of θ in Θ for which there is such a φ and let $\varphi(m, \theta)$ denote this φ with $B_\theta \subset f^{-1}(U_{\varphi(m,\theta)})$. Now

$$D = \bigcup_{m=1}^{\infty} \bigcup \left\{ B_\theta \setminus \text{int } f^{-1}(U_{\varphi(m,\theta)}) : \theta \in \Theta^{(m)} \right\},$$

and D, being a countable union of σ-discrete unions of closed sets, is an $\mathcal{F}_\sigma$-set. $\square$

Our next lemmas use a modification of a technique of Banach [2].

Lemma 2.12   *Let X be a metric space and let Y be an arcwise connected (metric) space. Let $\{F_\gamma : \gamma \in \Gamma\}$ be a discrete family of closed sets in X and let $\{y_\gamma : \gamma \in \Gamma\}$ be a set of points in Y. Then there is a continuous map $f : X \to Y$ with*

$$f(x) = y_\gamma, \quad \text{when } x \in F_\gamma,$$

for each γ in Γ.

Proof. Let ρ be the metric on X. (We do not make explicit use of the metric on Y.) Let y^* be any point in Y. For each γ in Γ, choose an arc joining y_γ to y^* in Y, that is a continuous function

$$\varphi_\gamma : [0, 1] \to Y,$$

with

$$\varphi_\gamma(0) = y_\gamma \text{ and } \varphi_\gamma(1) = y^*.$$

Note that in the important special case when Y is a convex set in a normed linear space, we simply take

$$\varphi_\gamma(t) = y_\gamma + t(y^* - y_\gamma), \quad 0 \le t \le 1.$$

For each closed set F in X we write

$$d(x, F) = \inf\{\rho(x,f) : f \in F\}.$$

Then $d(x, F)$ satisfies the Lipschitz condition

$$|d(x, F) - d(\xi, F)| \le \rho(x, \xi)$$

for all x, ξ in X. Further $d(x, F) = 0$, if and only if $x \in F$.

We make use of some particular cases of this function. Write

$$d(x) = d\left(x, \bigcup\{F_\gamma : \gamma \in \Gamma\}\right),$$

$$d_\gamma(x) = d(x, F_\gamma), \quad \gamma \in \Gamma,$$

$$e_\gamma(x) = d\left(x, \bigcup\{F_\beta : \beta \in \Gamma, \beta \ne \gamma\}\right), \quad \gamma \in \Gamma.$$

Note that the unions used in the definitions of $d(x)$ and of $e_\gamma(x)$ are closed sets since $\{F_\gamma : \gamma \in \Gamma\}$ is a discrete family of closed sets. Note also that, when $\alpha, \beta \in \Gamma$ and $\alpha \ne \beta$, then

$$d(x) \le e_\alpha(x) \le d_\beta(x)$$

for all x.

For each γ in Γ introduce the set

$$G_\gamma = \left\{x : d_\gamma(x) < \frac{1}{3}e_\gamma(x)\right\}.$$

Clearly G_γ is open and

$$F_\gamma \subset G_\gamma,$$

since $d_\gamma(x) = 0$ and $e_\gamma(x) \ge 0$ on F_γ. We verify that the family

$$\{G_\gamma : \gamma \in \Gamma\}$$

is discrete.

We first use a simple argument to show that the sets G_γ, $\gamma \in \Gamma$, are disjoint. Suppose that for some $\alpha, \beta \in \Gamma$ with $\alpha \ne \beta$, we have

$$x \in G_\alpha \cap G_\beta.$$

Then

$$d_\alpha(x) < \frac{1}{3}e_\alpha(x) \le \frac{1}{3}d_\beta(x),$$

$$d_\beta(x) < \frac{1}{3}e_\beta(x) \le \frac{1}{3}d_\alpha(x).$$

Adding these inequalities

$$d_\alpha(x) + d_\beta(x) < \frac{1}{3}\Big(d_\alpha(x) + d_\beta(x)\Big).$$

Hence

$$d_\alpha(x) = d_\beta(x) = 0$$

and

$$x \in F_\alpha \cap F_\beta$$

contrary to our hypotheses. Thus the sets G_γ, $\gamma \in \Gamma$ are disjoint.

We use a refinement of this argument to show that the family $\{G_\gamma : \gamma \in \Gamma\}$ is discrete. Consider any x^* in X. If $x^* \in F_{\gamma^*}$ for some γ^* in Γ then x^* has the open neighborhood G_{γ^*} that meets no set G_γ with $\gamma \ne \gamma^*$. So we may suppose that

$$x^* \notin \bigcup\{F_\gamma : \gamma \in \Gamma\}.$$

Then $d(x^*) > 0$. Consider the neighborhood U of x^* defined by

$$U = \Big\{x : \rho(x, x^*) < \frac{1}{2}d(x^*)\Big\}.$$

Suppose that for some α, β in Γ with $\alpha \ne \beta$, the set U meets G_α at some point a and that U meets G_β at b. Then

$$d_\alpha(a) < \frac{1}{3}e_\alpha(a) \le \frac{1}{3}d_\beta(a),$$

$$d_\beta(b) < \frac{1}{3}e_\beta(b) \le \frac{1}{3}d_\alpha(b).$$

Adding these inequalities, we obtain

$$d_\alpha(a) + d_\beta(b) < \frac{1}{3}\Big(d_\beta(a) + d_\alpha(b)\Big).$$

Using the Lipschitz conditions, we have

$$d_\alpha(x^*) + d_\beta(x^*) - \rho(a, x^*) - \rho(b, x^*)$$

$$< \frac{1}{3}\Big(d_\alpha(x^*) + d_\beta(x^*)\Big) + \frac{1}{3}(\rho(a, x^*) + \rho(b, x^*)).$$

Thus

$$\frac{4}{3}d(x^*) \le \frac{2}{3}\left(d_\alpha(x^*) + d_\beta(x^*)\right)$$
$$< \frac{4}{3}(\rho(a, x^*) + \rho(b, x^*))$$
$$< \frac{4}{3}d(x^*),$$

since a and b belong to U. This is impossible. Consequently, $\{G_\gamma : \gamma \in \Gamma\}$ is a discrete family.

We now define the function g by taking

$$g(x) = \varphi_\gamma\left\{\frac{4d_\gamma(x)}{d_\gamma(x) + e_\gamma(x)}\right\}, \quad \text{for } x \in G_\gamma, \ \gamma \in \Gamma,$$

and

$$g(x) = y^*,$$

for

$$x \in F^* = X \backslash \bigcup\{G_\gamma : \gamma \in \Gamma\}.$$

Note that on G_γ, we have

$$d_\gamma(x) + e_\gamma(x) > 0,$$

since x cannot belong to the two sets

$$F_\gamma \quad \text{and} \quad \bigcup\{F_\beta : \beta \in \Gamma, \ \beta \ne \gamma\}.$$

Further

$$\frac{4d_\gamma(x)}{d_\gamma(x) + e_\gamma(x)} = 0$$

on F_γ, and on G_γ

$$0 \le \frac{4d_\gamma(x)}{d_\gamma(x) + e_\gamma(x)} < \frac{\frac{4}{3}e_\gamma(x)}{\frac{1}{3}e_\gamma(x) + e_\gamma(x)} = 1.$$

Thus g is welldefined on X, and

$$g(x) = y_\gamma \quad \text{on } F_\gamma \text{ for } \gamma \in \Gamma.$$

It remains to prove that g is continuous. If x^* is in the open set G_γ for some γ in Γ, then

$$g(x) = \varphi_\gamma\left(\frac{4d_\gamma(x)}{d_\gamma(x) + e_\gamma(x)}\right)$$

for all x sufficiently close to x^*. Thus g is continuous at x^*, if $x^* \notin F^*$. Now suppose that $x^* \in F^*$. Then x^* has a neighborhood, U say, that meets at most one of the sets G_γ. If U meets no such set, then g takes the constant value y^* on U. Otherwise, we have just one γ in Γ for which U meets G_γ. On U we have

$$g(x) = y^*,$$

or

$$x \in G_\gamma \text{ and } g(x) = \varphi_\gamma\left(\frac{4d_\gamma(x)}{d_\gamma(x) + e_\gamma(x)}\right).$$

Since $x^* \notin G_\gamma$, we have

$$d_\gamma(x^*) \geq \frac{1}{3}e_\gamma(x^*).$$

Thus

$$\frac{4d_\gamma(x)}{d_\gamma(x) + e_\gamma(x)} \to 1$$

and

$$g(x) \to y^*$$

as $x \to x^*$ through G_γ, and g is continuous at x^*, as required. $\square$

Lemma 2.13 *Let X be a metric space and let Y be an arcwise connected metric space. Then a function from X to Y that is constant on the sets of a disjoint discretely σ-decomposable family of $\mathcal{F}_\sigma$-sets covering X is of the first Baire class.*

Proof. Let f be constant on the disjoint discretely σ-decomposable family

$$\{U_\gamma : \gamma \in \Gamma\}$$

of $\mathcal{F}_\sigma$-sets covering X. By Lemma 2.2 we can write

$$U_\gamma = \bigcup_{i=1}^\infty F_\gamma^{(i)}, \quad \gamma \in \Gamma,$$

with

$$F_\gamma^{(1)} \subset F_\gamma^{(2)} \subset \cdots, \quad \gamma \in \Gamma,$$

and with

$$\{F_\gamma^{(i)} : \gamma \in \Gamma\}$$

a discrete family of closed sets, for $i \geq 1$.

Now let y_γ be the constant value taken by f on U_γ for $\gamma \in \Gamma$. By Lemma 2.10, there is a continuous map $f_i : X \to Y$ with

$$f_i(x) = y_\gamma \quad \text{when } x \in F_\gamma^{(i)},$$

for $i \geq 1$, $\gamma \in \Gamma$. As i increases, the sets $F_\gamma^{(i)}$ increase to fill out the whole set U_γ. Hence, for $x \in U_\gamma$ we have

$$f_i(x) = y_\gamma = f(x),$$

for all sufficiently large i. Thus the continuous functions f_i converge pointwise to f and f is of the first Baire class. $\square$

Lemma 2.14 *Let X be a metric space and let C be a convex set in a normed vector space Y. If a sequence of functions of the first Baire class, as functions from X to C, converges uniformly to a function from X to C, then the limit function is of the first Baire class as a function from X to C.*

Proof. Let f be a function from X to C that is the uniform limit of functions f_n, $n \geq 1$, from X to C that are of the first Baire class as functions from X to C. We use the supremum "norm"

$$\|g\| = \sup\{\|g(x)\| : x \in X\}$$

for functions from X to Y, allowing the "norm" to take the value $+\infty$. Replacing the uniformly convergent sequence f_n, $n \geq 1$, by a suitable subsequence of itself, we may suppose that

$$\|f_n - f\| < 2^{-n}, \quad \text{for } n \geq 1.$$

For each $n \geq 1$, we can choose a sequence h_{nm}, $m \geq 1$, of continuous functions from X to C converging pointwise to f_n.

We use induction to define an array

$$g_{nm}, \quad 1 \leq n \leq m,$$

of moderated continuous functions from X to C. We take

$$g_{1m} = h_{1m}, \quad \text{for } m \geq 1.$$

When g_{nm} has been defined for $m \geq n$, we define g_{n+1m} by

$$g_{n+1m}(x) = g_{nm}(x) + \{h_{n+1m}(x) - g_{nm}(x)\},$$

$$\text{if } \|h_{n+1m}(x) - g_{nm}(x)\| < 2^{-n};$$

$$g_{n+1m}(x) = g_{nm}(x) + \frac{2^{-n}}{\|h_{n+1m}(x) - g_{nm}(x)\|} \{h_{n+1m}(x) - g_{nm}(x)\},$$

$$\text{if } \|h_{n+1m}(x) - g_{nm}(x)\| \geq 2^{-n}.$$

Since the functions h_{nm} are all continuous functions from X to C, and C is convex, it follows inductively that the functions g_{nm} are all continuous functions from X to C. Further, if n and x are fixed, we verify that $g_{nm}(x) = h_{nm}(x)$ for all sufficiently large m. This holds for all m when $n = 1$. Suppose that $n \geq 1$ and that

$$g_{nm}(x) = h_{nm}(x), \quad \text{for } m \geq m(n).$$

Then, for all sufficiently large m, we have

$$\|h_{n+1m}(x) - g_{nm}(x)\| = \|h_{n+1m}(x) - h_{nm}(x)\|.$$

Since $h_{nm}(x)$, $h_{n+1m}(x)$ converge to $f_n(x)$ and $f_{n+1}(x)$ and

$$\|f_{n+1}(x) - f_n(x)\| < 2^{-n},$$

we have

$$\|h_{n+1m}(x) - g_{nm}(x)\| < 2^{-n},$$

and so

$$g_{n+1m}(x) = h_{n+1m}(x)$$

for all sufficiently large values of m.

Note also that the definition ensures that

$$\|g_{n+1m} - g_{nm}\| \leq 2^{-n}$$

for $m \geq n + 1 \geq 2$.

It remains to verify that the sequence

$$g_{nn}, \quad n \geq 1,$$

of continuous functions from X to C converges pointwise to f. Let x be fixed in X, and let $n \geq 1$ be given. Choose $m_0 > n$ so large that

$$g_{nm}(x) = h_{nm}(x), \quad \text{for } m \geq m_0,$$

and

$$\|h_{nm}(x) - f_n(x)\| < 2^{-n}, \quad \text{for } m \geq m_0.$$

Then, for $m \geq m_0$,

$$\|g_{nn}(x) - f(x)\|$$

$$\leq \|g_{nn}(x) - g_{nm}(x)\| + \|g_{nm}(x) - f_n(x)\| + \|f_n(x) - f(x)\|$$

$$= \|g_{nn}(x) - g_{nm}(x)\| + \|h_{nm}(x) - f_n(x)\| + \|f_n(x) - f(x)\|$$

$$\leq \sum_{r=n}^{m-1} \|g_{rn}(x) - g_{r+1n}(x)\| + 2^{-n} + 2^{-n}$$

$$< \sum_{r=n}^{\infty} 2^{-r} + 2^{-n+1} = 2^{-n+2}.$$

This yields the required pointwise convergence. $\square$

Lemma 2.15 *Let X be a metric space and let C be a convex set in a normed vector space Y. Then a σ-discrete function of the first Borel class from X to Y taking its values in C is of the first Baire class as a function from X to C.*

Proof. Let f be such a function from X to C. By Lemmas 2.6 and 2.8, f is the uniform limit of a sequence f_n, $n \geq 1$, of functions f_n, $n \geq 1$, of functions from X to C, each function being constant on the sets of a σ-decomposable partition of X into $\mathcal{F}_\sigma$-sets. By Lemma 2.13, each function f_n is of the first Baire class as a function from X to C. The result now follows from Lemma 2.14. $\square$

It will be convenient to summarize the main results of this chapter.

Theorem 2.1 *Let X and Y be metric spaces. Let f be a function from X to Y. The following conditions on f are equivalent:*

  (a) f is a σ-discrete function of the first Borel class;

  (b) f has a closed σ-discrete base;

  (c) f is the uniform limit of a sequence of functions f_1, f_2, ..., each being constant on the sets of a corresponding discretely σ-decomposable partition of X into $\mathcal{F}_\sigma$-sets.

  When these conditions are satisfied, the set of discontinuities of f is an $\mathcal{F}_\sigma$-set of the first Baire category in X. If, in addition, Y is arcwise connected, f is the uniform limit of a sequence of functions of the first Baire class. If, in addition, Y is a convex set in a normed linear space, f is of the first Baire class.

2.3 WHEN IS A FUNCTION OF THE FIRST BOREL CLASS ALSO OF THE FIRST BAIRE CLASS?

Lemma 2.15 gives a partial answer to this question. Recently M. Fosgerau [13] has given a much more satisfactory answer. We quote, without proof, one of his main results.

Theorem 2.2 (Fosgerau) *Let X and Y be metric spaces.*

  (i) If Y is arcwise connected and locally arcwise connected then each σ-discrete function of the first Borel class from X to Y is of the first Baire class.

(ii) *If Y is complete and each function from* [0, 1] *to Y that is of the first*
      Borel class is also of the first Baire class, then Y is arcwise connected
      and locally arcwise connected.

We remark that Hansell [19, Theorem 3], has shown that every Borel measurable function from [0, 1] to a metric space Y is necessarily σ-discrete. We are grateful to D. Preiss for permission to include the following example here. Note that the function f of the example is necessarily of the first Borel class without being of the first Baire class.

Example 2.1 *There is a function f from the unit interval* [0, 1] *to a compact arcwise connected set P in* $\mathbb{R}^2$ *that is the uniform limit of functions of the first Baire class in P but is not itself of the first Baire class in P.*

Construction   We take P to be the union in $\mathbb{R}^2$ of the line segment

$$[0, 1] \times \{1\}$$

with the product

$$C \times [0, 1]$$

of the Cantor ternary set C (taken in its original form) with the unit interval. It is easy to see that P is compact and arcwise connected (but it is not locally connected).

Each real t with $0 \leq t \leq 1$ has a unique binary decimal expansion

$$t = 0.(t_1)(t_2)(t_3)...,$$

where, when $t \geq 0$, the sequence of binary digits $t_1, t_2, t_3, ...$ is not ultimately zero. We take f to be the map that takes t in [0, 1] with such a representation to the point $(x(t), 0)$ of P where $x(t)$ is defined by the ternary decimal expansion

$$x(t) = 0.(2t_1)(2t_2)(2t_3)... .$$

The intersections with P of the open rectangles, with sides parallel to the coordinate axes, form a base for the open sets in P. Clearly, $f^{-1}(I)$, for any such rectangle I, is either empty or a general subinterval of I. Hence the closed intervals in I with rational end-points form a countable closed base for f. Since P is arcwise connected, it follows from Theorem 2.1, that f is the limit of a uniformly convergent sequence of functions of the first Baire class from [0, 1] to P. The reader should have no difficulty in exhibiting f explicitly as the uniform limit of functions that are exhibited as pointwise limits of continuous functions from [0, 1] to P.

It remains to prove that f is not itself of the first Baire class as a function from [0, 1] to P. We suppose that $\varphi_1, \varphi_2, ...$ is a sequence of continuous functions from [0, 1] to P that converges pointwise to f, and we seek a contradiction. Let

$$\varphi_r(t) = (\xi_r(t), \eta_r(t)), \quad 0 \leq t \leq 1.$$

Write

$$G_r = \left\{t : \eta_r(t) > \frac{1}{2}\right\}, \text{ and } H_r = \bigcup_{s=r}^{\infty} G_s.$$

Then H_r is an open set in $[0, 1]$. We prove that H_r is dense in $[0, 1]$ for each $r \geq 1$. Consider any interval (t_1, t_2) with $0 \leq t_1 < t_2 \leq 1$. Then

$$x(t_1) < x(t_2).$$

By the pointwise convergence of φ_p to f,

$$(\xi_p(t), \eta_p(t)) \to (x(t), 0)$$

as $p \to \infty$, for each t in $[0, 1]$. So we can choose $p > r$ so that

$$|\xi_p(t_i) - x(t_i)| < \frac{1}{3}(x(t_2) - x(t_1)), \quad i = 1, 2;$$

$$|\eta_p(t_i)| \leq \frac{1}{2}, \quad i = 1, 2.$$

This ensures that

$$\xi_p(t_1) \neq \xi_p(t_2)$$

and that $\varphi_p(t_1)$, $\varphi_p(t_2)$ lie on distinct line segments of the form

$$\{\xi_p(t_1)\} \times \left[0, \frac{1}{2}\right],$$

$$\{\xi_p(t_2)\} \times \left[0, \frac{1}{2}\right],$$

with $\xi_p(t_1)$, $\xi_p(t_2)$ in C. Since φ_p is continuous, and since

$$C \setminus ([0, 1] \times \{1\})$$

is not connected between $\varphi_p(t_1)$, $\varphi_p(t_2)$, it follows that $\eta_p(t) = 1$ for some t with $t_1 < t < t_2$. This shows that H_r is dense in $[0, 1]$ for $r \geq 1$. Now

$$\bigcap_{r=1}^{\infty} H_r,$$

being the intersection of dense open sets, is a dense G_δ-set in $[0, 1]$. Now, if τ belongs to this nonempty set, we have

$$\eta_p(\tau) \geq \frac{1}{2}$$

for infinitely many values of p, and $\varphi_p(\tau)$ cannot converge to $f(\tau)$.

2.4 REMARKS

The results in section 2.1 may be found in Hansell [19]. A space X is called *subparacompact* if each open cover of X has a σ-discrete closed refinement. Although we have chosen to work with a metric space X, all the results of section 2.1 hold when X is a subparacompact space all of whose closed sets are G_δ-sets.

The results in section 2.2 up to and including Lemma 2.11 are either from Hansell [19] or are obtained by the methods he used there. Again these results hold for a subparacompact space all of whose closed sets are G_δ-sets. Lemma 2.12 and its sequels depend on the assumption that X be metric.

We should re-emphasize that σ-discrete functions between metric spaces are rather common. Hansell [19] shows that every function from a metric space to a separable metric space is σ-discrete and, more importantly, every Borel function from a space X that is a Souslin-$\mathcal{F}$ set in some complete metric space $\hat{X}$, to a metric space Y, is σ-discrete. Further, each pointwise limit of a sequence of σ-discrete functions is σ-discrete, Hansell [19].

Note that the conclusion in Lemma 2.11 that the set of points of discontinuity of f is of the first Baire category in X tells us little about the nature of the set of points of continuity of f, unless we know that X is of the second Baire category in itself. We give a simple example of a σ-discrete function of the first Borel class with no points of continuity.

Let f map 0 and the rational numbers of the form $2p/q$, with $q \geq 1$ and q odd, to 0 and the other rational numbers to 1. Then f is a function from $\mathbb{Q}$ to $\mathbb{R}$ that is σ-discrete and of the first Borel class with no point of continuity.

The technique for proving Lemma 2.14 is like that used in proving the Tietze extension theorem.

Lemma 2.15 is due to Hansell [19] who uses a much less direct method.

Chapter 3

Selectors for upper semi-continuous functions with nonempty compact values

In this chapter we obtain selectors of the first Borel class and of the first Baire class for certain upper semi-continuous set-valued functions with nonempty compact values.

Before we state the main results it will be convenient to give some definitions. We suppose that Z is a space with a Hausdorff topology τ and also a metric d, not necessarily related to the topology τ. We say that (Z, τ) is *fragmented down to* $\epsilon > 0$ *by* d, if each nonempty subset of Z has a nonempty τ-relatively open subset of d-diameter less than ϵ. We say that (Z, τ) is *fragmented by* d, if it is fragmented down to ϵ by d, for each $\epsilon > 0$. We say that (Z, τ) is *σ-fragmented down to* ϵ *by* d, if Z can be expressed in the form

$$Z = \bigcup_{n=1}^{\infty} Z_n,$$

with each set Z_n fragmented down to ϵ by d for each $n \geq 1$. We say that (Z, τ) is *σ-fragmented by* d, if (Z, τ) is σ-fragmented by d for each $\epsilon > 0$. In the cases of the σ-fragmentation definitions we add the clause "using sets from S" if the sets $Z_n, n \geq 1$, can be chosen from the family S.

Considerable effort has been devoted to the fragmentation properties of Banach spaces (see, e.g., [25, 31, 33–41]). We quote some of these results later.

We obtain the following results in section 3.2.

Theorem 3.1 *Let F be an upper semi-continuous set-valued function, from a metric space X to a metric space Y, taking only nonempty compact values. Then F has a selector f that is σ-discrete and of the first Borel class and the set of points of discontinuity of f is an $\mathcal{F}_\sigma$-set of the first Baire category in X. If Y is arcwise connected, the selector f will be the uniform limit of a sequence of functions of the first Baire class. If, in addition, Y is a convex set in a normed linear space, then f is of the first Baire class.*

Theorem 3.2 *Let X be a metric space and let K be a weakly closed set in a Banach space Y (perhaps $K = Y$). Suppose that each bounded subset of*

(K, weak) *is σ-fragmented by the norm using weakly closed sets. If F is an upper semi-continuous set-valued function from X to (K, weak) taking only nonempty weakly compact values, then F has a selector $f : X \longrightarrow (K, \text{norm})$ that is of the first Baire class as a map from X to the weakly closed convex hull of K, with the norm topology and has an $\mathcal{F}_\sigma$-set of the first Baire category as its set of discontinuities.*

We remark that if a Banach space Y is weakly compactly generated then (Y, weak) is σ-fragmented by the norm using weakly closed sets [35, Theorem 2.1(c)].

Theorem 3.3 *Let X be a metric space and let K^* be a convex weak* closed set in the dual Y^* of a Banach space Y (perhaps $K^* = Y^*$). Suppose that (K^*, weak^*) is σ-fragmented by the norm using weak* closed sets. If F is any upper semi-continuous set-valued function from X to (K^*, weak^*) taking only nonempty weak* compact values, then F has a selector $f : X \longrightarrow (K^*, \text{norm})$ that is of the first Baire class and has an $\mathcal{F}_\sigma$-set of the first Baire category as its set of points of discontinuity.*

Note that the dual space Y^* has the Radon–Nikodým property if and only if the unit ball (B^*, weak^*) of Y^* is fragmented by the norm. In this case (since $Y^* = \bigcup_{n=1}^{\infty} nB^*$), it follows that (Y^*, weak^*) is σ-fragmented by the norm using weak* closed sets.

We obtain these results in section 3.2 as simple consequences of the following abstract result that is a modified version of part of the result of Hansell, Jayne and Talagrand [22, Theorem 1$'$].

Theorem 3.4 *Let X be a metric space. Let Z be a Hausdorff space with topology τ and with a metric d defining a topology on Z at least as strong as τ. Suppose that (Z, τ) is σ-fragmented by d using τ-closed sets. If F is an upper semi-continuous set-valued map from X to (Z, τ) taking only nonempty compact values, then F has a selector $f : X \longrightarrow (Z, d)$ that is σ-discrete and of the first Borel class and that has an $\mathcal{F}_\sigma$-set of the first Baire category in X as its set of points of discontinuity.*

See Remark 1 below for a (modified) version of the second part of the theorem of Hansell, Jayne and Talagrand. The first stage of the proof of Theorem 3.4 uses a reduction of the upper semi-continuous set-valued function F, taking only nonempty compact values, to a minimal function H of this type with its values contained in those of F. In section 3.3 we develop some of the properties of such minimal functions and their selectors.

3.1 A GENERAL THEOREM

Throughout this section we suppose that X is a metric space and that Z is a space with Hausdorff topology τ and a metric d defining a topology on Z at least as strong as the topology τ. We use $\mathbb{F}$ to denote the set of upper semi-continuous set-valued maps from X to (Z, τ) taking only nonempty compact values.

A set valued map H in $\mathbb{F}$ will be said to be *minimal in* $\mathbb{F}$, or simply *minimal*, if there is no L in $\mathbb{F}$ with

$$L(x) \in H(x), \qquad \text{for all } x \text{ in } X,$$

and

$$L(x_0) \neq H(x_0), \qquad \text{for some } x_0 \text{ in } X.$$

Lemma 3.1 *If F is an element of $\mathbb{F}$, then there is an element H of $\mathbb{F}$, that is minimal in $\mathbb{F}$ and that has*

$$H(x) \subset F(x), \qquad \text{for all } x \text{ in } X.$$

Proof. Let $\mathcal{H}$ be the family of all elements H in $\mathbb{F}$ with

$$H(x) \subset F(x), \qquad \text{for all } x \text{ in } X.$$

Of course $F \in \mathcal{H}$. We introduce a partial order on $\mathcal{H}$ by writing

$$H_1 \leq H_2,$$

whenever H_1, H_2 belong to $\mathcal{H}$ and

$$H_1(x) \subset H_2(x), \qquad \text{for all } x \text{ in } X.$$

Suppose that

$$\mathcal{H}_A = \{H_\alpha : \alpha \in A\}$$

is a subset of $\mathcal{H}$ indexed by the set A and totally ordered by "$\leq$". Consider the set-valued function H defined by

$$H(x) = \bigcap \{H_\alpha(x) : \alpha \in A\}, \qquad \text{for all } x \text{ in } X.$$

For fixed x, the family $\{H_\alpha : \alpha \in A\}$ is a nested family of nonempty compact sets. Thus H takes only nonempty compact values.

Now consider any closed set K in Z. We have

$$H(x) \cap K \neq \emptyset,$$

if and only if

$$\bigcap\{H_\alpha \cap K : \alpha \in A\} \neq \emptyset;$$

that is, if and only if

$$H_\alpha \cap K \neq \emptyset, \quad \text{for all } \alpha \text{ in } A;$$

that is, if and only if

$$x \in \bigcap\{\{\hat{x} : H_\alpha(\hat{x}) \cap K \neq \emptyset\} : \alpha \in A\}.$$

But each set

$$\{\hat{x} : H_\alpha(\hat{x}) \cap K \neq \emptyset\}$$

is closed in X by the semi-continuity of H_α, for each $\alpha \in A$. Hence the set

$$\{x : H(x) \cap K \neq \emptyset\}$$

is closed in X and H is upper semi-continuous. Thus $H \in \mathcal{H}$ and

$$H \leq H_\alpha, \quad \text{for all } \alpha \text{ in } A.$$

Now Zorn's lemma shows that there is an element H that is minimal in $\mathbb{F}$ with $H \leq F$ as required. $\square$

Lemma 3.2 *Let H be a minimal element of $\mathbb{F}$.*

(a) *Let K be a closed set in (Z, τ) and let G be an open set in X. If $H(x) \cap K \neq \emptyset$ for each x in G, then $H(x) \subset K$ for all x in G.*

(b) *Let U be an open set in (Z, τ) and let V be an open set in X with $H(V) \cap U \neq \emptyset$. Then there is a nonempty open set G in X with*

$$G \subset V \quad and \quad H(G) \subset U.$$

Proof.

(a) It is easy to verify that the set-valued function defined by

$$L(x) = H(x), \quad \text{if } x \notin G,$$

$$L(x) = H(x) \cap K, \quad \text{if } x \in G,$$

is an upper semi-continuous set-valued map from X to (Z, τ), taking only nonempty compact values, with

$$L(x) \subset H(x)$$

for each x in X. Since H is minimal in $\mathbb{F}$, L must coincide with H, ensuring that

$$F(x) \subset K$$

for each x in G.

(b) Now suppose that U is an open set in (Z, τ) and V is an open set in X with $H(V) \cap U \neq \emptyset$. Suppose that we have

$$H(x) \setminus U \neq \emptyset$$

for each x in V. By part (a) this would yield

$$H(x) \subset Z \setminus U, \quad \text{for all } x \text{ in } V,$$

and we could not have had $H(V) \cap U \neq \emptyset$. Hence there will be a point, x_0 say, in V with

$$H(x_0) \subset U.$$

By the semi-continuity of H, we can choose an open set G in X with

$$x_0 \in G \subset V$$

and

$$H(G) \subset U$$

as required. $\quad \square$

In the next lemmas we use $\mathbb{F}(X, Z)$ to denote the family of all upper semi-continuous set-valued functions from the metric space X to the Hausdorff space Z, taking only nonempty compact values.

Lemma 3.3 *Suppose that for some $\epsilon > 0$, the space (Z, τ) is fragmented down to ϵ by the metric d. If F is any element of $\mathbb{F}(X, Z)$, there is a disjoint discretely σ-decomposable family $\{O_\alpha : \alpha \in A\}$ of $\mathcal{F}_\sigma$-sets and a set-valued function $H : X \to Z$ with:*

(1) $H(x) \subset F(x)$ for each x in X;

(2) d-diam $H(O_\alpha) < \epsilon$ for each α in A;

(3) H restricted to O_α belonging to $\mathbb{F}(O_\alpha, Z)$ for each α in A.

Proof. By Lemma 3.1 we may choose a minimal member H_0 of $\mathbb{F}$ with

$$H_0(x) \subset F(x) \quad \text{for each } x \text{ in } X.$$

Since (Z, τ) is fragmented down to ϵ by d, and

$$H_0(X) \neq \emptyset,$$

we can choose an open set U in (Z, τ) with

$$H_0(X) \cap U \neq \emptyset$$

and

$$d\text{-diam}(H_0(X) \cap U) < \epsilon.$$

By Lemma 3.2 (b), we can choose a nonempty open set G_0 in X with

$$H_0(G_0) \subset U.$$

Now

$$H_0(G_0) \subset H_0(X) \cap U,$$

and so

$$d\text{-diam}(H_0(G_0)) < \epsilon.$$

We take

$$F_0 = X \quad \text{and} \quad F_1 = X \setminus G_0.$$

Now F_0 and F_1 are closed sets and:

$$F_1 \supset F_0 \quad \text{and} \quad F_1 \neq F_0;$$

$$H_0 \quad \text{belongs to} \quad \mathbb{F}(F_0, Z);$$

$$F(x) \supset H_0(x) \quad \text{for } x \in F_0;$$

and

$$d\text{-diam}(H_0(F_0 \setminus F_1)) < \epsilon.$$

We now show that for some ordinal Γ, there is a sequence of closed sets

$$F_\gamma, \quad 0 \leq \gamma < \Gamma,$$

and of set-valued functions

$$H_\gamma : F_\gamma \to Z, \quad 0 \leq \gamma < \Gamma,$$

satisfying the following conditions:

(1) $F_0 = X$;

(2) $F_\alpha \supset F_\beta$ and $F_\alpha \neq F_\beta$ if $0 \leq \alpha < \beta < \Gamma$;

(3) H_γ belongs to $\mathbb{F}(F_\gamma, Z)$ for $0 \leq \gamma < \Gamma$;

(4) $F(x) \supset H_\alpha(x) \supset H_\beta(x)$ if $0 \leq \alpha \leq \beta < \Gamma$, and $x \in F_\beta$;

(5) $d\text{-diam } H_\gamma(F_\gamma \setminus F_{\gamma+1}) < \epsilon$ for $0 \leq \gamma < \gamma + 1 < \Gamma$;

(6) $F_\lambda = \bigcap \{F_\alpha : 0 \leq \alpha < \lambda\}$ if λ is a limit ordinal;

(7) $\bigcap \{F_\alpha : 0 \leq \alpha < \Gamma\} = \emptyset$.

Later it will be convenient to write

$$O_\gamma = F_\gamma \setminus F_{\gamma+1}, \quad \text{if } 0 \leq \gamma < \gamma + 1 < \Gamma.$$

Note that the conditions (1)–(5) are already satisfied in as far as they apply to

$$H_0, F_0 \quad \text{and} \quad F_1.$$

Suppose that $\delta + 1 \geq 1$ is a successor ordinal and that F_α, $0 \leq \alpha \leq \delta + 1$, and H_α, $0 \leq \alpha \leq \delta$, have been defined satisfying the conditions (1)–(5) for the values of α, β and γ for which these conditions make sense. If $F_{\delta+1} = \emptyset$ we terminate the process, taking $\Gamma = \delta + 2$. Otherwise we apply the argument of the first part of this proof to the restriction of H_δ to $F_{\delta+1}$ to choose a closed set $F_{\delta+2}$ strictly contained in $F_{\delta+1}$ and a set-valued function $H_{\delta+1} : F_{\delta+1} \to Z$ in $\mathbb{F}(F_{\delta+1}, Z)$ with

$$H_{\delta+1}(x) \subset H_\delta(x), \quad \text{for } x \text{ in } F_{\delta+1},$$

and with

$$d\text{-diam } H_{\delta+1}(F_{\delta+1} \setminus F_{\delta+2}) < \epsilon.$$

In this case the conditions (1)–(5) are satisfied by

$$F_\alpha, \; 0 \leq \alpha \leq \delta + 2 \quad \text{and} \quad H_\alpha, \; 0 \leq \alpha \leq \delta + 1.$$

Now suppose that δ is a limit ordinal and that F_α and H_α have been defined for $0 \leq \alpha < \delta$ satisfying the conditions (1)–(6) for α, β, γ and λ less than δ. If

$$\bigcap \{F_\gamma : 0 \leq \alpha < \delta\} = \emptyset,$$

we take $F_\delta = \emptyset$ and $\Gamma = \delta + 1$ and terminate the construction. Otherwise we take

$$F_\delta = \bigcap \{F_\gamma : 0 \leq \gamma < \delta\}$$

and we define a set-function $K_\delta : F_\delta \to Z$, by

$$K_\delta(x) = \bigcap \{F_\gamma(x) : 0 \leq \gamma < \delta\}, \quad \text{for } x \in F_\delta.$$

As in the proof of Lemma 3.1, it is easy to verify that $K_\delta \in \mathbb{F}(F_\delta, Z)$. By the first part of this lemma, applied to K_δ on F_δ, we can choose a closed set $F_{\delta+1}$ strictly contained in F_δ and a set-valued function $H_{\delta+1} : F_\delta \to Z$ in $\mathbb{F}(F_\delta, Z)$ with

$$H_\delta(x) \subset H_\gamma(x), \quad \text{for } 0 \leq \gamma < \delta \text{ and } x \in F_\delta$$

and

$$d\text{-diam } H_\delta(F_\delta \setminus F_{\delta+1}) < \epsilon.$$

In this case the conditions (1)–(6) are satisfied by

$$F_\alpha, \quad 0 \leq \alpha \leq \delta + 1 \quad \text{and} \quad H_\alpha, \quad 0 \leq \alpha \leq \delta.$$

The construction must terminate, with $F_{\Gamma-1} = \emptyset$, for some Γ, since the sets F_γ, $0 \leq \gamma$, are strictly decreasing. Thus the construction is completed by transfinite induction and the conditions (1)–(7) are satisfied.

We now write

$$O_\gamma = F_\gamma \setminus F_{\gamma+1}, \quad \text{if } 0 \le \gamma < \gamma+1 < \Gamma,$$

and define the set-valued function $H : X \to Z$ by

$$H(x) = H_\gamma(x), \quad \text{if } x \in O_\gamma, \; 0 \le \gamma < \gamma+1 < \Gamma.$$

Note that the family

$$\{O_\gamma : 0 \le \gamma < \gamma+1 < \Gamma\}$$

is a disjoint partition of X into $\mathcal{F}_\sigma$-sets.

By Lemma 2.1 this partition is discretely σ-decomposable. The conditions of our lemma are satisfied by taking A to be the set of ordinals γ with $0 \le \gamma < \gamma+1 < \Gamma$. $\square$

Lemma 3.4 *Suppose that for some $\epsilon > 0$, the space (Z, τ) is σ-fragmented down to ϵ by d using τ-closed sets. If F is any element of $\mathbb{F}(X, Z)$, there is a disjoint discretely σ-decomposable family $\{O_\alpha : \alpha \in A\}$ of $\mathcal{F}_\sigma$-sets and a set-valued function $H : X \to Z$ satisfying the conditions (1), (2) and (3) of Lemma 3.3.*

Proof. Since (Z, τ) is σ-fragmented down to ϵ by d using τ-closed sets,

$$Z = \bigcup \{Z^{(n)} : n \ge 1\},$$

with each set $Z^{(n)}, n \ge 1$, fragmented down to ϵ by d and τ-closed. For each $n \ge 1$, write

$$X^{(n)} = \{x : F(x) \cap Z^{(n)} \ne \emptyset\}.$$

Since F is upper semi-continuous, $X^{(n)}$ is a closed subset of X, and the set function

$$F^{(n)} : X^{(n)} \to Z^{(n)}$$

defined by

$$F^{(n)} = F(x) \cap Z^{(n)}, \; x \in X^{(n)}$$

belongs to $\mathbb{F}(X^{(n)}, Z^{(n)})$. Applying Lemma 3.3 to this map $F^{(n)}$, we obtain a disjoint discretely σ-decomposable family $\{O_\alpha^{(n)} : \alpha \in A^{(n)}\}$ of $\mathcal{F}_\sigma$-sets in $X^{(n)}$, and a set-valued function $H^{(n)}$ from $X^{(n)}$ to $Z^{(n)}$ satisfying:

(1) $H^{(n)}(x) \subset F^{(n)}(x)$ for each x in $X^{(n)}$;

(2) d-diam $H^{(n)}(O_\alpha^{(n)}) < \epsilon$ for each $\alpha \in A^{(n)}$; and

(3) $H^{(n)}$ restricted to $O_\alpha^{(n)}$ belongs to $\mathbb{F}(O_\alpha^{(n)}, Z^{(n)})$ for each $\alpha \in A^{(n)}$.

Let $X^{(0)} = \emptyset$. Since $F(x) = \emptyset$ for each x in X the sets

$$X^{(n)} \setminus X^{(n-1)}, \; n \ge 1,$$

form a countable disjoint partition of X into $\mathcal{F}_\sigma$-sets. Now the family of sets

$$\{O_\alpha^{(n)} \cap (X^{(n)} \setminus X^{(n-1)}) : \alpha \in A^{(n)},\ n \geq 1\}$$

forms a disjoint discretely σ-decomposable partition of X into $\mathcal{F}_\sigma$-sets.

Define the set-valued function $H : X \longrightarrow Z$ by taking

$$H(x) = H^{(n)}(x), \quad \text{when} \quad x \in X^{(n)} \setminus X^{(n-1)},\ n \geq 1.$$

It is easy to verify that this set function H together with the above partition of X satisfies the conditions (1), (2) and (3) of this lemma. $\square$

Proof of Theorem 3.4. We are supposing that X is a metric space and that Z is a space with a Hausdorff topology τ and also a metric d defining a topology on Z at least as strong as τ. We also suppose that (Z, τ) is σ-fragmented by d using τ-closed sets.

Consider any upper semi-continuous set-valued map F from X to (Z, τ), taking only nonempty τ-compact values. Using Lemma 3.4 we can choose a disjoint discretely σ-decomposable partition $\{O_\alpha^{(1)} : \alpha \in A^{(1)}\}$ of X by $\mathcal{F}_\sigma$-sets and set-valued function $H^{(1)} : X \longrightarrow Z$ satisfying the conditions:

(1) $H^{(1)}(x) \subset F(x)$ for each x in X;

(2) d-diam $H^{(1)}(O_\alpha^{(1)}) < \frac{1}{2}$ for each $\alpha \in A^{(1)}$; and

(3) $H^{(1)}$ restricted to $O_\alpha^{(1)}$ belongs to $\mathbb{F}(O_\alpha^{(1)}, Z)$ for each $\alpha \in A^{(1)}$.

Applying Lemma 3.4 to each set function $H^{(1)}$ restricted to each $O_\alpha^{(1)}$, with $\alpha \in A^{(1)}$, we can find a disjoint discretely σ-decomposable family

$$\{O_{\alpha\beta}^{(2)} : \beta \in B_\alpha^{(2)}\}$$

of $\mathcal{F}_\sigma$-sets partitioning $O_\alpha^{(1)}$ and a set-valued function $H_\alpha^{(2)} : O_\alpha^{(1)} \longrightarrow Z$ satisfying the conditions:

(1) $H_\alpha^{(2)}(x) \subset H^{(1)}(x)$ for each $x \in O_\alpha$;

(2) d-diam $H_\alpha^{(2)}(O_{\alpha\beta}^{(2)}) < \frac{1}{4}$ for each $\beta \in B_\alpha^{(2)}$; and

(3) $H_\alpha^{(2)}$ restricted to $O_{\alpha\beta}^{(2)}$ belongs to $\mathbb{F}(O_{\alpha\beta}^{(2)}, Z)$ for each $\beta \in B_\alpha^{(2)}$.

By Lemma 2.3, the family

$$\{O_{\alpha\beta}^{(2)} : \alpha \in A^{(1)},\ \beta \in B_\alpha^{(2)}\},$$

which we denote by

$$\{O_\alpha^{(2)} : \alpha \in A^{(2)}\},$$

with

$$A^{(2)} = \{\alpha\beta : \alpha \in A^{(1)}, \ \beta \in B_\alpha^{(2)}\},$$

is a disjoint discretely σ-decomposable family of $\mathcal{F}_\sigma$-sets partitioning X. Further defining $H^{(2)} : X \to Z$ by

$$H^{(2)}(x) = H_\alpha^{(2)}(x), \quad \text{if } x \in O_\alpha^{(2)}$$

(using the new notation) we find that:

(1) $H^{(2)}(x) \subset H^{(1)}(x)$ for each $x \in X$;

(2) d-diam $H_\alpha^{(2)}(O_\alpha^{(2)}) < \frac{1}{4}$ for each $\alpha \in A^{(2)}$; and

(3) $H^{(2)}$ restricted to $O_\alpha^{(2)}$ belongs to $\mathbb{F}(O_\alpha^{(2)}, Z)$ for each $\beta \in A^{(2)}$.

Iterating this process, we build a sequence of discretely σ-decomposable partitions $\{O_\alpha^{(n)} : \alpha \in A^{(n)}\}$, $n \geq 1$, of X into $\mathcal{F}_\sigma$-sets and a sequence $H^{(n)}$, $n \geq 1$, of set-valued functions from X to Z satisfying the conditions:

(1) $F(x) \supset H^{(1)}(x) \supset H^{(2)}(x) \supset \cdots$ for each $x \in X$;

(2) d-diam $H_\alpha^{(n)}(O_\alpha^{(n)}) < 2^{-n}$ for $n \geq 1$, $\alpha \in A^{(n)}$;

(3) $H(n)$ restricted to $O_\alpha^{(n)}$ belongs to $\mathbb{F}(O_\alpha^{(n)}, Z)$ for $n \geq 1$, $\alpha \in A^{(n)}$; and

(4) $\{O_\alpha^{(n+1)} : \alpha \in A^{(n+1)}\}$ refines $\{O_\alpha^{(n)} : \alpha \in A^{(n)}\}$ for $n \geq 1$.

Now, for each x in X, the sequence

$$F(x), \ H^{(1)}(x), \ H^{(2)}(x), \ldots$$

is a decreasing sequence of nonempty τ-compact sets, with

$$d\text{-diam} H^{(n)}(x) < 2^{-n}, \quad n \geq 1.$$

Thus

$$\bigcap_{n=1}^\infty H(n)(x)$$

is a nonempty set consisting of a single point of X. Let $h(x)$ denote this point.

In this way we have constructed a selector h for F. It remains to determine the Borel nature of h. For each $n \geq 1$, we construct a function $h^{(n)}$ from X to Z. For each α in $A^{(n)}$ we choose a fixed point $z_\alpha^{(n)}$ in $H^{(n)}(O_\alpha^{(n)})$. We define $h^{(n)}$ by taking

$$h^{(n)}(x) = z_\alpha^{(n)}, \quad \text{for } x \in O_\alpha^{(n)}, \ \alpha \in A^{(n)}.$$

Thus $h^{(n)}$ is constant on each of the sets of a discretely σ-decomposable partition of X into $\mathcal{F}_\sigma$-sets. Since $h(x)$ and $h^{(n)}(x)$ both lie in $H^{(n)}(O_\alpha^{(n)})$ when $x \in O_\alpha^{(n)}$, we have

$$d\Big(h(x), h^{(n)}(x)\Big) < 2^{-n},$$

for $x \in X$ and $n \geq 1$.

Thus h is the uniform limit in (Z, d) of the sequence of functions $h^{(n)}$, each constant on the sets of a discretely σ-decomposable partition of X into $\mathcal{F}_\sigma$-sets. By Theorem 2.1, h regarded as a function to (Z, d) is a selector for F and is σ-discrete and of the first Borel class and the set of its points of discontinuity is an $\mathcal{F}_\sigma$-set of the first Baire category in X. $\square$

3.2 SPECIAL THEOREMS

In this section we prove Theorems 3.1, 3.2 and 3.3, stated in the introduction to this chapter.

Proof of Theorem 3.1. We are supposing that F is an upper semi-continuous set-valued function, from a metric space X to a metric space Y, taking only nonempty compact values. We wish to apply Theorem 3.4 to X, the space Z taken to be Y with its metric topology τ generated by its metric d, and to F. Note that Z is trivially σ-fragmented by using the set Z. So we can apply Theorem 3.4 and obtain a selector f for F that is σ-discrete and of the first Borel class and that has an $\mathcal{F}_\sigma$-set of the first Baire category in X as its set of points of discontinuity. Further, using Theorem 2.1, if Y is arcwise connected, f is the uniform limit of a sequence of functions of the first Baire class. If, in addition, Y is a convex set in a normed linear space, f is itself of the first Baire class. $\square$

Proof of Theorem 3.2. We are supposing that X is a metric space and that K is a weakly closed set in a Banach space Y. Also (K, weak) is σ-fragmented by the norm using weakly closed sets. Consider a set-valued function $F : X \to (K, \text{weak})$ that is upper semi-continuous and takes only nonempty weakly compact values. By Theorem 3.4, F has a selector

$$f : (K, \text{norm}) \to X$$

that is σ-discrete and of the first Borel class and its set of points of discontinuity is an $\mathcal{F}_\sigma$-set of the first Baire category in X. By Theorem 2.1, f is of the first Baire class as a map from X to the weakly closed convex hull of K, with the norm topology. $\square$

Proof of Theorem 3.3. This result follows from Theorems 3.4 and 2.1, just as Theorem 3.2 follows from these two theorems, by using the weak* topology in the dual Banach space Y^*. $\square$

3.3 MINIMAL UPPER SEMI-CONTINUOUS SET-VALUED MAPS

In this section we obtain some properties of set-valued functions that are minimal amongst the upper semi-continuous set-valued functions from a metric space X to a Hausdorff space Z that take only nonempty compact values.

To state the theorem that we prove, we need to recall the definition of a lower semi-continuous metric on a topological space. A metric on a topological space (Z, τ) is said to be *lower semi-continuous* if, for each $r > 0$, the set

$$\{(z, \zeta) : d(z, \zeta) \leq r\}$$

is closed in $Z \times Z$. Note that the norm metric in a Banach space with its weak topology is always lower semi-continuous, as is the norm metric on a dual Banach space with its weak* topology.

Theorem 3.5 *Let (Z, τ) be a Hausdorff space. Let d be a metric on Z defining a topology at least as strong as τ. Let $F : X \rightarrow (Z, \tau)$ be a set valued function that is upper semi-continuous and takes only nonempty compact values, and is minimal amongst all such maps. Let P be the set of points x of X for which $F(x)$ is a single point. Then each selector f for F is τ-continuous precisely at the points of P. If the metric d is lower semi-continuous then there is a set Q contained in P with the property that each selector f for F is d-continuous precisely at the points of Q. If, in addition, (Z, τ) is fragmented by d, then Q is a countable intersection of dense open sets.*

Note that when X is not the union of nowhere dense sets, then the last clause ensures that Q, and so also P, is nonempty and dense in X. It will be convenient to base the proof on a series of lemmas. In these lemmas, we take (Z, τ) to be a Hausdorff space with a metric d defining a topology on Z at least as strong as τ. We also assume that $F : X \rightarrow (Z, \tau)$ is minimal amongst the upper semi-continuous set-valued functions from X to (Z, τ) taking only nonempty compact values.

Lemma 3.5 *If p belongs to P, then each selector f for F is τ-continuous at p.*

Proof. Let U be any open set in (Z, τ) containing

$$\{f(p)\} = F(p).$$

Since F is upper semi-continuous, there is an open set G in X containing p, with

$$F(G) \subset U.$$

Thus

$$f(x) \in F(G) \subset U,$$

for all x in G. Hence f is τ-continuous at p. $\square$

Lemma 3.6 *Let f be a selector for F that is τ-continuous at a point p of X. Then $p \in P$.*

Proof. Suppose that $F(p)$ is not a singleton. Then $F(p)$ contains $f(p)$ and also some other point, z say. Since (Z, τ) is a Hausdorff space we can choose disjoint τ-open sets U and V with $f(p) \in V$ and $z \in U$. Since f is τ-continuous at p, we can choose an open set G with $p \in G$ and

$$f(x) \in V, \quad \text{for all } x \in G.$$

By Lemma 3.2 (a) this implies that

$$F(p) \subset K$$

despite the fact that $z \notin K$. This shows that $F(p)$ is a singleton, so that $p \in P$.

We now introduce the set Q. For each $\epsilon > 0$ write

$$Q_\epsilon = \bigcup \{U : U \text{ is open in } X \text{ with } d\text{-diam } F(U) < \epsilon\}$$

and take

$$Q = \bigcap_{n=1}^{\infty} Q_{1/n}. \quad \square$$

Lemma 3.7 *$Q \subset P$ and if $q \in Q$ each selector f for F is d-continuous at q.*

Proof. Suppose that $q \in Q$. We can choose a sequence $U_n, n \geq 1$, of open sets in X, containing q with

$$d\text{-diam } F(U_n) < 1/n.$$

First this implies that

$$d\text{-diam } F(q) = 0,$$

so that $F(q)$ is a single point. Thus $Q \subset P$.

Secondly, we conclude that F, regarded as a set-valued map from X to (Z, d) is upper semi-continuous at q, ensuring that each selector f for F is d-continuous at q. $\quad \square$

Lemma 3.8 *Suppose that the metric d is lower semi-continuous on (Z, τ). Let f be a selector for F that is d-continuous at q. Then $q \in Q$.*

Proof. Consider any $\epsilon > 0$. Since d is lower semi-continuous the set

$$K = \{z : d(z, f(q)) \leq \epsilon\}$$

is τ-closed. The set

$$K_0 = \{z : d(z, f(q)) < \epsilon\}$$

is d-open. Since f is d-continuous at q, we can choose an open set G in X containing q with

$$f(x) \in K_0 \subset K, \quad \text{for each } x \text{ in } G.$$

Thus

$$F(x) \cap K \neq \emptyset, \quad \text{for each } x \text{ in } G.$$

By Lemma 3.2 (a)

$$F(x) \subset K,$$

for all x in G. Hence

$$d\text{-diam}F(G) < 3\epsilon,$$

ensuring that $q \in Q_{3\epsilon}$. Since $\epsilon > 0$ is arbitrary this yields $q \in Q$. $\quad\square$

Lemma 3.9 *If (Z, τ) is fragmented by d, the set Q is the countable intersection of open sets dense in X.*

Proof. Consider any $\epsilon > 0$. Let G be nonempty open set in X. Then $F(G) \neq \emptyset$. Since (Z, τ) is fragmented by d, we can choose an open set V in (Z, τ) with

$$F(G) \cap V \neq \emptyset$$

and

$$d\text{-diam}(F(G) \cap V) < \epsilon.$$

Suppose that we had

$$F(x) \cap (Z \setminus V) \neq \emptyset$$

for each x in G. Then by Lemma 3.2 (a) we would have

$$F(x) \subset Z \setminus V, \quad \text{for all } x \text{ in } G,$$

contrary to the choice of V with

$$F(G) \cap V \neq \emptyset.$$

Thus

$$F(x) \subset V$$

for some x, say x_0, in G. By the upper semi-continuity of F at x_0, we can choose an open neighborhood W of x_0 contained in G with

$$F(W) \subset V.$$

Now

$$d\text{-diam}F(W) \le d\text{-diam}(F(G) \cap V) < \epsilon.$$

Thus

$$x_0 \in W \subset Q_\epsilon \quad \text{and} \quad x_0 \in G.$$

This shows that Q_ϵ is dense in X. The result follows. $\square$

Proof of Theorem 3.5. The properties of P follow from Lemmas 3.5 and 3.6. The properties of Q and its relations with P and with X follow from Lemmas 3.7, 3.8 and 3.9. $\square$

3.4 REMARKS

1. Theorems 3.1 and 3.2 are simplified and modified forms of parts of the results of Hansell, Jayne and Talagrand [22] (see their Theorem $10'$ and its corollaries). Proofs of variants of Theorems 3.2 and 3.3 are also given in [31]. In their paper Hansell, Jayne and Talagrand [22] obtain a number of other results and in particular prove the following result in a more abstract form (see their Corollary 2).

   Theorem 3.6 (Hansell, Jayne and Talagrand) *Let K be a compact Hausdorff space. Let $C(K)$ denote the Banach space of continuous functions on K with the supremum norm and let $C_p(K)$ denote the space $C(K)$ taken with the topology of pointwise convergence. Let $F : X \longrightarrow C_p(K)$ be an upper semi-continuous set-valued function from a complete metric space X to $C_p(k)$ taking only nonempty compact values. Then F has a selector of the first norm Borel class which is σ-discrete, whose set of points of norm continuity is a dense G_δ-subset of X, and which is the norm pointwise limit of sequence of norm continuous functions from X to $C(K)$.*

   Their proof is rather complicated and we do not present it in this book. However, using a method of V. V. Srivatsa we do give (a complicated) proof of an even more powerful result (see Theorem 6.3 below).

2. The selectors obtained in Theorems 3.1–3.4 are specified as σ-discrete functions of the first Borel class. Although we have not emphasized this, by Theorem 2.1, they have other interesting properties.

3. Theorem 3.5 is a simplified version of part of a result recently obtained by Jayne, Namioka and Rogers [35] (see their Theorem 3.1). It is related to the work of Kenderov [45].

4. The concept of a Hausdorff space (Z, τ) that is fragmented by a metric d on Z, was introduced in [31, see section 2], in order to have a uniform terminology for a number of well-known Banach space situations, in particular, (a) a Banach space with its weak topology and its norm

metric, having the point of continuity property and (b) a dual Banach space with its weak* topology and its norm metric having the Radon–Nikodým property. The term "fragmentable" or "fragments" was used since, for each $\epsilon > 0$, the concept enabled the space Z to be decomposed into the union

$$\bigcup \{Z_\gamma : 0 \le \gamma < \Gamma\}$$

of disjoint "fragments" Z_γ, $0 \le \gamma < \Gamma$, with:

(1) $\bigcup \{Z_\gamma : 0 \le \alpha < \gamma\}$ τ-open for each γ with $0 \le \gamma < \Gamma$;

(2) Z_γ is the difference between two τ-open sets, for $0 \le \gamma < \Gamma$;

(3) d-diam $Z_\gamma < \epsilon$ for $0 \le \gamma < \Gamma$.

5. The concept of σ-fragmentability was introduced by Jayne, Namioka and Rogers [36]. It is easy to check that an equivalent definition of the σ-fragmentation of a Hausdorff space (Z, τ) by a metric d on Z is:

(Z, τ) *is σ-fragmented by d if and only if, for each cover*

$$C = \{C_\alpha : \alpha \in A\}$$

of Z by d-open sets, it is possible to write

$$Z = \bigcup_{n=1}^{\infty} Z_n,$$

with each Z_n, $n \ge 1$, having the property that each nonempty subset of Z_n has a nonempty relatively τ-open subset contained in some set belonging to C.

It is clear (by considering the covers C_ϵ of Z by all d-open sets of d-diameter less than ϵ) that if the new definition is satisfied then so can the old. On the other hand, if the original definition is satisfied and the d-open cover C is given, one can first split Z as

$$Z = \bigcup_{m=1}^{\infty} Z^{(m)},$$

with

$$Z^{(m)} = \{z : B(z, 1/m) \subset C \quad \text{for some } C \in C\},$$

and then use the fact that each $Z^{(m)}$ is σ-fragmented down to $1/(3m)$, for $m \ge 1$, to split up Z further. It is clear from this new definition that the question of whether or not (Z, τ) is σ-fragmented by d, depends only on the relationship between the topology τ and the metric topology determined by d, rather than the actual metric d. Recently Namioka and Pol [62] (see their Theorem 4.3), have established deeper results, proving, in particular, that a Banach space with its weak topology, say (Y, weak), is σ-fragmented by its norm if and only if (Y, weak) is an "almost Čech-analytic space",which is a property of (Y, weak) purely as a topological space, without reference to any metric topology or linear structure.

6. For a detailed discussion of the structure of upper semi-continuous set-valued maps from one metric space to another, see [32].

7. Consider the following question: "Let Y be a Banach space with its weak topology or a dual Banach space with its weak* topology. If F is an upper semi-continuous set-valued function from a metric space to Y taking only nonempty compact values, is it possible to find a complete metric space $\hat{X}$ containing X and an extension $\hat{F}$ of F to an upper semi-continuous set-valued function from $\hat{X}$ to Y taking only nonempty compact values?" If the answer to this question was always "Yes!", then there would be no essential loss of generality in taking X in Theorems 3.2 and 3.3 to be a complete metric space. Further, the theorems could then end with the more elegant claim that the selectors f were continuous at the points of a dense G_δ-set in X. As it happens, the answer to the question is "Not always!". We give an example to justify this answer.

Example *There is an upper semi-continuous set-valued map from the rational numbers $\mathbb{Q}$ to the Banach space c_0 with its weak topology, taking only nonempty norm compact values, that has no upper semi-continuous extension from any completion of $\mathbb{Q}$ to (c_0, weak) taking nonempty values.*

Construction For convenience, we identify $\mathbb{Q}$ with the set $\mathbb{Q}_0$ of all rational numbers r with $0 < r < 1$. Each such r has precisely two continued fraction expansions

$$r = \frac{1}{a_1+} \frac{1}{a_2+} \cdots \frac{1}{a_n},$$

$$r = \frac{1}{a_1+} \frac{1}{a_2+} \cdots \frac{1}{(a_{n-1})+} \frac{1}{1},$$

with $a_i \geq 1$ for $1 \leq i < n$ and $a_n \geq 2$. Let $x(r)$ be the point in c_0 defined by taking $x_i(r) = 1$ if i takes one of the n values

$$a_1,$$

$$a_1 + a_2,$$

$$\vdots$$

$$a_1 + a_2 + \cdots + a_n$$

and $x_i(r) = 0$, otherwise. Let $y(r)$ be the point in c_0 defined by taking $y_i(r) = 1$ if i takes one of the $n + 1$ values

$$a_1,$$

$$a_1 + a_2,$$

$$\vdots$$

$$a_1 + a_2 + \cdots + (a_n - 1),$$

$$a_1 + a_2 + \cdots + a_n,$$

and $y_i(r) = 0$ otherwise.

Note that $x(r)$ and $y(r)$ differ just because

$$x_{a_1 + a_2 + \cdots + (a_n - 1)}(r) = 0,$$

$$y_{a_1 + a_2 + \cdots + (a_n - 1)}(r) = 1.$$

Write

$$F(r) = \{x(r), y(r)\},$$

for all r in $\mathbb{Q}_0$. Then F is clearly a set-valued function from $\mathbb{Q}_0$ to c_0 taking only nonempty norm compact values. We verify F is upper semi-continuous for the weak topology of c_0. For a fixed r in $\mathbb{Q}_0$ let G be a weak open set in c_0 containing $F(r)$. Then we can choose basic weak open sets G_x, G_y in c_0 with

$$x(r) \in G_x \subset G,$$

$$y(r) \in G_y \subset G.$$

We may suppose that these basic open sets in c_0 are of the form

$$G_x = \{\xi : \langle \xi - x(r), x_i^* \rangle < \epsilon,\ 1 \le i \le N\},$$

$$G_y = \{\xi : \langle \xi - y(r), y_i^* \rangle < \epsilon,\ 1 \le i \le M\}$$

where $\epsilon > 0$, and x_i^*, $1 \le i \le N$, y_i^*, $1 \le i \le M$, are points in the dual space l_1 of c_0, with

$$\|x_i^*\|_1 = 1,\ 1 \le i \le N,$$

$$\|y_i^*\|_1 = 1,\ 1 \le i \le M.$$

We can choose L so large that

$$|\langle \xi, x_i^* \rangle| < \epsilon,\ 1 \le i \le N,$$

$$|\langle \xi, y_i^* \rangle| < \epsilon,\ 1 \le i \le M,$$

for all ξ in c_0 with $\|\xi\|_\infty = 1$ and

$$\xi_i = 0, \quad \text{for } 1 \le i \le L.$$

We now study $F(s)$ for rational s very close to r. We use the standard continued fraction notation and formulae; see the first four pages of the section on continued fraction in Hardy and Wright's book on the Theory of Numbers [23]. Thus $p_0 = 0$, $q_0 = 1$ and

$$\frac{p_1}{q_1}, \frac{p_2}{q_2}, \ldots, \frac{p_n}{q_n} = r$$

are the convergents for

$$r = \frac{1}{a_1+} \frac{1}{a_2+} \cdots \frac{1}{a_n}.$$

Consider any rational number s with

$$0 < |s - r| \le \frac{1}{(L+2)q_n^2}.$$

First suppose that

$$(-1)^n(s - r) > 0,$$

so that

$$0 < (-1)^n(s - r) \le \frac{1}{(L+2)q_n^2}.$$

Then there is a unique rational a'_{n+1} satisfying

$$(-1)^n(s - r) = \frac{1}{q_n(a'_{n+1}q_n + q_{n-1})}.$$

Solving for a'_{n+1}, and using $q_{n-1} < q_n$,

$$a'_{n+1} = \frac{(-1)^n}{q_n^2(s - r)} - \frac{q_{n-1}}{q_n} > L + 2 - 1 = L + 1.$$

Solving now for s, and using

$$p_nq_{n-1} - p_{n-1}q_n = (-1)^{n-1},$$

we have

$$s = \frac{p_n}{q_n} + \frac{(-1)^n}{q_n(a'_{n+1} + q_{n-1})} = \frac{a'_{n+1}p_n + p_{n-1}}{a'_{n+1}q_n + q_{n-1}}.$$

Using the formulae

$$p_{n+1} = a_{n+1}p_n + p_{n-1}, \quad q_{n+1} = a_{n+1}q_n + q_{n-1},$$

we have

$$s = \frac{1}{a_1+} \frac{1}{a_2+} \cdots \frac{1}{a_n+} \frac{1}{a'_{n+1}},$$

with $a'_{n+1} > L + 1$, so that s has a complete continued fraction expansion of the form

$$s = \frac{1}{a_1+} \frac{1}{a_2+} \cdots \frac{1}{a_n+} \frac{1}{b_{n+1}+} \cdots \frac{1}{b_m},$$

with $m \geq n + 1, b_{n+1} \geq L + 1, b_m \geq 2$. This ensures that, when i satisfies

$$1 \leq i < a_1 + a_2 + \cdots + a_n + b_{n+1} - 1,$$

we have

$$x_i(s) = y_i(s) = x_i(r).$$

Since

$$a_1 + a_2 + \cdots + a_n + b_{n+1} - 1 \geq b_{n+1} \geq L + 1,$$

our choice of L ensures that

$$F(s) = \{x(s), y(s)\} \subset G_x.$$

Now consider the case when

$$(-1)^{n+1}(s - r) > 0.$$

Applying the same arguments to the continued fraction

$$r = \frac{1}{a_1+} \frac{1}{a_2+} \cdots \frac{1}{(a_n - 1)+} \frac{1}{1},$$

but using the formula

$$p_{n+1}q_n - p_n q_{n+1} = (-1)^n,$$

we find that

$$F(s) \subset G_y.$$

Thus, in each case, $F(s) \subset G$. This shows that F is upper semi-continuous.

We note that r and s are two distinct rational numbers between 0 and 1, and that the four points

$$x(r), \quad y(r), \quad x(s), \quad y(s)$$

are all distinct. Since all the coordinates are either 0 or 1, these four points from a regular tetrahedron of side 1 in (c_0, norm). This in itself shows that, if f is any selector for F, then f can be norm continuous at no point of Q_0.

Now suppose that $\hat{\mathbb{Q}}_0$ is the completion of $\mathbb{Q}_0$ under any metric that is consistent with its topology. Let $\hat{F}$ be any extension of F to $\hat{\mathbb{Q}}$ taking only nonempty sets in c_0 as its values. We suppose that $\hat{F}$ is upper semi-continuous to (c_0,weak) and seek a contradiction. By Srivatsa's Theorem, see Theorem 6.2 below, $\hat{F}$ will have a selector $\hat{f}$ of the first Baire class on $\hat{\mathbb{Q}}_0$. Since $\hat{\mathbb{Q}}_0$ is complete, the function $\hat{f}$ will be continuous at some point t of $\hat{\mathbb{Q}}_0$. So t will have a neighborhood, U say, with

$$\text{diam } \hat{f}(U) < \frac{1}{2}.$$

Since $\hat{\mathbb{Q}}_0$ is a completion of $\hat{\mathbb{Q}}_0$, we can find distinct rationals r and s in U. Since

$$\|\hat{f}(r) - \hat{f}(s)\|_\infty = 1,$$

this yields the required contradiction, showing that this example satisfies the statement.

Chapter 4

Selectors for compact sets

In this chapter we derive some results, not dissimilar from those in Chapter 3, from a different point of view. We combine the method of Jayne and Rogers [31] with methods developed by Ghoussoub, Maurey and Schachermayer [14] in a paper containing many interesting selection results.

Recall that the *Hausdorff metric d* on the space of nonempty bounded closed sets of a metric space (X, ρ) is defined by

$$d(H, K) = \inf\{\epsilon > 0 : H \subset K_\epsilon \text{ and } K \subset H_\epsilon\},$$

where

$$H_\epsilon = \{x : \rho(x, H) < \epsilon\}$$

is the *ϵ-neighborhood* of H.

Before we state the main results in this chapter, we state a very simple result that sets the pattern for the subsequent results.

Theorem 4.1 *Let $\mathcal{K}$ be the space of nonempty closed sets in the unit square Q, taken with the Hausdorff metric. Then there is a function $s : \mathcal{K} \to Q$ of the first Borel class satisfying the following conditions:*

(a) $s(K) \in K$ for each K in $\mathcal{K}$;

(b) if $K_1 \subset K_2$ and $s(K_2) \in K_1$ then $s(K_1) = s(K_2)$;

(c) if $K_1, K_2, \ldots$ is a decreasing sequence of closed sets in $\mathcal{K}$ converging to a set $\mathcal{K}_0$ in $\mathcal{K}$, then $s(K_i)$ converges to $s(K)$ as $i \to \infty$.

The simple proof of this result is given in section 4.1 below. The proof makes the conditions (a), (b) and (c) appear natural. We now state the main results of this chapter.

Theorem 4.2 *Let (Y, ρ) be a metric space. Let $\mathcal{K}_\rho$ be the space of non-empty closed subsets of Y with the Hausdorff metric derived from ρ.*

Then there is a function $s : \mathcal{K}_\rho \to Y$ that is σ-discrete and of the first Borel class satisfying the following conditions:

(a) $s(K) \in K$ for each K in $\mathcal{K}_\rho$;

(b) if $K_1 \subset K_2$ and $s(K_2) \in K_1$ then $s(K_1) = s(K_2)$;

(c) if $\{K_\alpha : \alpha \in A\}$, with A a directed set, is a decreasing net of nonempty compact sets with

$$K = \bigcap \{K_\alpha : \alpha \in A\},$$

then

$$s(K_\alpha) \longrightarrow s(K)$$

following A;

(d) if F is any upper semi-continuous set-valued function, from a metric space X to Y, taking only nonempty compact values, then $s \circ F$ is a selector for F that is σ-discrete and of the first Borel class.

A Banach space Y is said to be weakly σ-fragmented using weakly closed sets if, for each $\epsilon > 0$, it is possible to write

$$Y = \bigcup_{i=1}^{\infty} Y_i,$$

where each set Y_i is weakly closed and, for each nonempty subset S of Y_i there is a weak open set U in Y with

$$S \cap U \neq \emptyset, \quad \operatorname{diam} S \cap U < \epsilon.$$

Jayne, Namioka and Rogers [35] show that each weakly compactly generated Banach space is σ-fragmented in this way; furthermore each transfinite ℓ^p-sum ($1 \leq p < \infty$) of weakly compactly generated Banach spaces is σ-fragmented. They also show [38] that, if a Banach space has a norming extended Markashevich basis, then it is σ-fragmented using weakly closed sets.

Theorem 4.3 *Let Y be a Banach space that is weakly σ-fragmented using weakly closed sets. Let $\mathcal{K}_{\|\ \|}$ be the space of nonempty weakly compact subsets of Y with the Hausdorff metric derived from the norm.*

Then there is a function $s : \mathcal{K}_{\|\ \|} \longrightarrow (Y, norm)$ that is σ-discrete and of the first Baire class satisfying the following conditions:

(a) $s(K) \in K$ for each K in $\mathcal{K}_{\|\ \|}$;

(b) if $K_1 \subset K_2$ and $s(K_2) \in K_1$ then $s(K_1) = s(K_2)$;

(c) if $\{K_\alpha : \alpha \in A\}$, with A a directed set, is a decreasing net of nonempty weakly compact sets with

$$K = \bigcap \{K_\alpha : \alpha \in A\},$$

then

$$s(K_\alpha) \longrightarrow s(K), \quad \text{in norm,}$$

following A;

(d) *if F is any upper semi-continuous set-valued function, from a metric space X to Y with its weak topology, taking only nonempty weakly compact values, than s ∘ F is a selector for F that is, when regarded as a function from X to Y with its norm topology, σ-discrete and of the first Baire class.*

Weak* σ-fragmentation for a dual Banach space is defined similarly to the weak σ-fragmentation for a Banach space. If Y^* is the dual of an Asplund space Y, then Y^* is weak* σ-fragmented using weak* closed sets, see the remark after the statement of Theorem 3.3 above.

Theorem 4.4 *Let Y^* be a dual Banach space that is weak* σ-fragmented using weak* closed sets. Let $\mathcal{K}_{\|\ \|}$ be the space of nonempty weak* compact subsets of Y^* with the Hausdorff metric derived from the norm. Then there is a function $s : \mathcal{K}_{\|\ \|} \longrightarrow (Y^*, norm)$ that is σ-discrete and of the first Baire class satisfying the following conditions:*

(a) *$s(K) \in K$ for each K in $\mathcal{K}_{\|\ \|}$;*

(b) *if $K_1 \subset K_2$ and $s(K_2) \in K_1$ then $s(K_1) = s(K_2)$;*

(c) *if $\{K_\alpha : \alpha \in A\}$, with A a directed set, is a decreasing net of nonempty weak* compact sets with*

$$K = \bigcap \{K_\alpha : \alpha \in A\},$$

then

$$s(K_\alpha) \longrightarrow s(K), \quad \text{in norm,}$$

following A;

(d) *if F is any upper semi-continuous set-valued function, from a metric space X to Y^* with its weak* topology, taking only nonempty weakly compact values, then s ∘ F is a selector for F that is, when regarded as a function from X to Y^* with its norm topology, σ-discrete and of the first Baire class.*

4.1 A SPECIAL THEOREM

In this section we prove the simple Theorem 4.1 stated in the introduction to this chapter.

Proof of Theorem 4.1. Let $p : [0, 1] \longrightarrow Q$ be a Peano curve mapping $[0, 1]$ continuously onto the unit square Q. For each K in $\mathcal{K}, p^{-1}(K)$ is a closed set in

[0, 1]. So we may take $t(K)$ to be the smallest real number in $p^{-1}(K)$. Now take $s(K)$ to be $p(t(K))$ for each K in $\mathcal{K}$. Clearly $s(K) \in K$ for each K in $\mathcal{K}$.

Now if $K_1 \subset K_2$ then $p^{-1}(K_1) \subset p^{-1}(K_2)$, so that $t(K_2) \leq t(K_1)$. If, in addition, $s(K_2) \in K_1$ then $p(t(K_2)) = s(K_2) \in K_1$ and so $t(K_1) \leq t(K_2)$. Thus $t(K_1) = t(K_2)$ and so $s(K_1) = s(K_2)$.

Suppose that $K_1, K_2, \ldots$ is a decreasing sequence of closed sets in Q, converging in $\mathcal{K}$ to a closed set K_0. The corresponding sequence of real numbers

$$t(K_i), \quad i = 1, 2, \ldots,$$

is increasing and so tends to a limit in [0, 1], say t_*. If we had

$$t(K_0) < t(K_i)$$

for some $i \geq 1$, then K_0 would contain the point $s(K_0) = p(t(K_0))$ which is not in K_i. Hence

$$t(K_0) \geq \sup t(K_i) = t_*.$$

Since p is continuous,

$$p(t(K_i)) \rightarrow p(t_*) \quad \text{as } i \rightarrow \infty.$$

We have now proved that s satisfies the conditions (a), (b) and (c). It remains to show that s is of the first Borel class as a map from $\mathcal{K}$ to Q.

As a first step we show that, if F is any nonempty closed set in Q, then the set

$$\mathcal{K}(F) = \{K \in \mathcal{K} : K \cap F \neq \emptyset\}$$

is closed in $\mathcal{K}$. Consider any sequence $K_1, K_2, \ldots$ of sets of $\mathcal{K}(F)$ that converges in $\mathcal{K}$ to some closed set K_0. If K_0 does not meet F, then there is a positive distance, say 2ϵ, between F and K_0. So the ϵ-neighborhood $(K_0)_\epsilon$ of K_0 does not meet F. However, for all sufficiently large i, the set K_i lies in $(K_0)_\epsilon$ and cannot meet F. This contradiction shows that K_0 meets F. Hence $\mathcal{K}(F)$ is closed in $\mathcal{K}$.

Now consider any open set G in Q. To prove that s as a map from $\mathcal{K}$ to Q is of the first Borel class, we need to prove that $S^{-1}(G)$ is always an $\mathcal{F}_\sigma$-set in $\mathcal{K}$ with the Hausdorff metric. Consider the set

$$p^{-1}(G).$$

This set is open in [0, 1] and so is the union of a countable sequence of disjoint open intervals, say

$$p^{-1}(G) = \bigcup_{i=1}^{\infty} I_i,$$

so that

$$G = \bigcup_{i=1}^{\infty} p(I_i).$$

Thus

$$s^{-1}(G) = \bigcup_{i=1}^{\infty} s^{-1}(p(I_i)).$$

It now suffices to prove that each set

$$s^{-1}(p(I_i)), \quad i \geq 1,$$

is an $\mathcal{F}_\sigma$-set in $\mathcal{K}$. Now

$$K \in s^{-1}(p(I_i)),$$

if and only if

$$t(K) \in I_i.$$

If I_i is the interval $l_i < t < u_i$, the condition that $t(K)$ belongs to I_i reduces to the condition that

(i) $K \cap p([0, l_i]) = \emptyset$

and

(ii) $K((l_i, u_i)) \neq \emptyset.$

Since $p([0, l_i])$ is a closed set in Q, the result of the last paragraph shows that (i) restricts K to lie in an open set in the metric space $\mathcal{K}$, and so to lie in an $\mathcal{F}_\sigma$-set in $\mathcal{K}$. Since $p((l_i, u_i))$ is a countable union of closed sets in Q, the same result from the last paragraph shows that condition (ii) also restricts K to lie in an $\mathcal{F}_\sigma$-set. Since the intersection of two $\mathcal{F}_\sigma$-sets is also an $\mathcal{F}_\sigma$-set, the set $s^{-1}(p(I_i))$ is an $\mathcal{F}_\sigma$-set in $\mathcal{K}$. Similar considerations apply if I_i happens to be of the form $[0, u_i)$ or $(l_i, 1]$. Thus $s^{-1}(G)$ is an $\mathcal{F}_\sigma$-set in $\mathcal{K}$ and s is of the first Borel class. $\square$

4.2 A GENERAL THEOREM

We formulate a more general theorem that leads easily to the proofs of Theorems 4.2, 4.3 and 4.4. We first need to introduce some *ad hoc* terminology.

When a metric ρ is defined on a Hausdorff space Z, we say that ρ has the ϵ-*neighborhood property for compact sets* if whenever G is an open set in Z containing a compact set K of Z, the ϵ-neighborhood

$$K_\epsilon = \{z : \rho(z, k) < \epsilon\}$$

is contained in G for some $\epsilon > 0$.

Theorem 4.5 *Let Z be a completely regular Hausdorff space. Let ρ be a metric on Z with the ϵ-neighborhood property for compact sets, and suppose that closed sets in Z are ρ-closed. Suppose that Z is σ-fragmented by ρ using closed sets. Let $\mathcal{K}_\rho$ be the space of nonempty compact sets of Z with the Hausdorff metric derived from ρ.*

Then there is a function $s : \mathcal{K}_\rho \to (Z, \rho)$ that is σ-discrete and of the first Borel class satisfying the following conditions:

(a) *$s(K) \in K$ for each K in $\mathcal{K}_\rho$;*

(b) *if $K_1 \subset K_2$ and $s(K_2) \in K_1$ then $s(K_1) = s(K_2)$;*

(c) *if $\{K_\alpha : \alpha \in A\}$, with A a directed set, is a decreasing net of nonempty compact sets in Z with*

$$K = \bigcap \{K_\alpha : \alpha \in A\},$$

then

$$\rho(s(K_\alpha), s(K)) \to 0$$

following A;

(d) *if F is any upper semi-continuous set-valued function, from a metric space X to Z, taking only nonempty compact values, then $s \circ F$ is a selector for F that is, when regarded as a function from X to (Z, ρ), σ-discrete and of the first Borel class.*

We first establish a simple lemma showing that, if a locally convex topology on a normed vector space is weaker than the norm topology, then the norm metric has the ϵ-neighborhood property for the compact sets in the locally convex topology. For the theory of such topologies on normed vector spaces, see, for example, Kelley and Namioka [44].

Lemma 4.1 *Let $\mathcal{T}$ be a locally convex topology on a normed vector space Y, and suppose that all $\mathcal{T}$ open sets of Y are norm open. Then the norm metric has the ϵ-neighborhood property for Y with the topology $\mathcal{T}$.*

Proof. Let K be a $\mathcal{T}$ compact set contained in a $\mathcal{T}$ open set G. For each point k of K we can find a $\mathcal{T}$ open set $U(k)$, containing the origin 0, with

$$k + U(k) + U(k) \subset G.$$

The sets

$$k + U(k), \quad k \in K,$$

form a $\mathcal{T}$ open cover of the $\mathcal{T}$ compact set K. So we may choose a finite set $k_1, k_2, ..., k_n$ from K with

$$K \subset \bigcup_{i=1}^{n} \{k_i + U(k_i)\}.$$

Now

$$\bigcap_{i=1}^{n} U(k_i)$$

is a $\mathcal{T}$ open set containing 0. Hence we can choose $\epsilon > 0$ so that

$$B(0; \epsilon) \subset \bigcap_{i=1}^{n} U(k_i).$$

Now, if

$$h \in K_\epsilon = \{y : \min\{\|y - k\| : k \in K\} < \epsilon\},$$

we have

$$\|h - k\| < \epsilon$$

for some k in K. Then for some i, $1 \le i \le n$,

$$k \in k_i + U(k_i).$$

Thus

$$h = k_i + (h - k) + (k - k_i)$$

$$\in k_i + B(0; \epsilon) + U(k_i)$$

$$\subset k_i + U(k_i) + U(k_i) \subset G$$

and $K_\epsilon \subset G$, as required. $\square$

Before starting our next lemma, it is convenient to introduce another term from the theory of σ-fragmentability. When ρ is a metric on a Hausdorff space Z, we say that a set S in Z is *fragmented by ρ down to ϵ*, if, for each nonempty subset T of S, there is an open subset U in Z with

$$T \cap U \ne \emptyset \quad \text{and} \quad \rho\text{-diam}\,(T \cap U) < \epsilon.$$

Lemma 4.2 *Let Z be a completely regular Hausdorff space and let ρ be a metric on Z. Let Z_1, Z_0 be closed sets in Z with $D = Z_1 \backslash Z_0 \ne \emptyset$. Suppose that $\epsilon > 0$ and that Z_1 is fragmented by ρ down to ϵ. Then it is possible to choose a disjoint transfinite sequence*

$$\{B_\gamma : 0 \le \gamma < \Gamma\}$$

of $\mathcal{F}_\sigma$-sets in Z, such that, for $0 \le \gamma < \Gamma$ we have:

(a) *the union*

$$\bigcup_{0 \le \beta \le \gamma} B_\beta$$

is relatively open in D;

(b) B_γ *is nonempty with*

$$\rho\text{-diam}\,(\text{cl}\,B_\gamma) < \epsilon;$$

(c)

$$\bigcup_{0 \le \gamma < \Gamma} B_\gamma = D.$$

Proof. First consider any nonempty relatively closed set F of $D = Z_1 \backslash Z_0$. Then

$$F = \hat{F} \backslash Z_0$$

for some closed set $\hat{F}$ contained in Z_1. Since Z_1 is fragmented to ϵ, we can choose an open set U in Z with

$$F \cap U \ne \emptyset \quad \text{and} \quad \rho\text{-diam}\,(F \cap U) < \epsilon.$$

Choose a point z_0 of $F \cap U$. Then $z_0 \in U$, but $z_0 \notin (Z\backslash U) \cup Z_0$. Since Z is completely regular, we can choose a continuous function h on Z with

$$h(z_0) = 0 \quad \text{and} \quad h(z) = 1 \quad \text{on } (Z\backslash U) \cup Z_0.$$

Let F' be the set of all points of $\hat{F}$ with $h(z) < \frac{1}{2}$. The closure $\text{cl}\,F'$ is contained in the set of points of $\hat{F}$ with $h(z) \le \frac{1}{2}$. Hence $\text{cl}\,F'$ does not meet the set $(Z\backslash U) \cup Z_0$. Thus

$$\text{cl}\,F' \subset \hat{F}\backslash Z_0 = F$$

and

$$\text{cl}\,F' \subset U,$$

so that

$$\rho\text{-diam}\,(\text{cl}\,F') \le \rho\text{-diam}\,(F \cap U) < \epsilon.$$

Note that F' is an $\mathcal{F}_\sigma$-set contained in F and containing z_0. Further, since F' is relatively open in $\hat{F}$, the set $\hat{F}\backslash F'$ is relatively closed in $\hat{F}$ and

$$F\backslash F' = (\hat{F}\backslash Z_0)\backslash F'$$

$$= (\hat{F}\backslash F')\backslash Z_0$$

is relatively closed in $D = Z_1\backslash Z_0$.

We define the sequence $\{B_\gamma : 0 \leq \gamma < \Gamma\}$, together with a second sequence $\{R_\gamma : 0 \leq \gamma < \Gamma\}$ inductively to satisfy the conditions (a) and (b) and also the condition

(d)
$$R_\gamma = D \setminus \bigcup_{0 \leq \beta < \gamma} B_\beta.$$

We start by taking $R_0 = D$. By the result of the last paragraph, we can take B_0 to be a nonempty $\mathcal{F}_\sigma$-set contained in D, with ρ-diam (cl B_0) $< \epsilon$ and with $R_0 \setminus B_0$ relatively closed in D. Thus B_0 is relatively open in D.

Now suppose that for some ordinal $\gamma > 0$, the sets B_β, R_β have been defined, for $0 \leq \beta < \gamma$, satisfying the conditions (a), (b) and (d). Then

$$\bigcup_{0 \leq \beta < \gamma} B_\beta = \bigcup_{0 \leq \beta < \gamma} \left\{ \bigcup_{0 \leq \alpha \leq \beta} B_\alpha \right\}$$

is relatively open in D. If this union coincides with Z, we take $\Gamma = \gamma$ and the construction terminates. Otherwise, the set defined by

$$R_\gamma = D \setminus \bigcup_{0 \leq \beta < \gamma} B_\beta.$$

is relatively closed in D and nonempty. By the first paragraph of this proof, we can take B_γ to be a nonempty $\mathcal{F}_\sigma$-set in Z, contained in R_γ, with ρ-diam (cl B_γ) $< \epsilon$ and with $R_\gamma \setminus B_\gamma$ relatively closed in D. Now

$$\bigcup_{0 \leq \beta \leq \gamma} B_\beta = \left\{ \bigcup_{0 \leq \beta < \gamma} B_\beta \right\} \cup B_\gamma$$

$$= D \setminus \{R_\gamma \setminus B_\gamma\}$$

is relatively open in D, and the conditions (a), (b) and (d) are satisfied for this ordinal γ. As the sequence of sets

$$\bigcup_{0 \leq \beta < \gamma} B_\beta, \quad 0 \leq \gamma,$$

is strictly increasing, until the whole of D is covered, this transfinite process terminates with a cover of D.

Note that, since

$$B_\gamma \subset R_\gamma = D \setminus \bigcup_{0 \leq \beta < \gamma} B_\beta,$$

the sets B_γ, $0 \leq \gamma < \Gamma$, are disjoint. $\square$

Lemma 4.3 *Let Z be a completely regular Hausdorff space and let ρ be a metric on Z. Suppose that Z is σ-fragmented by ρ using closed sets. Let $\epsilon > 0$. Then it is possible to choose an index set Θ, a family*

$$\{S(\theta) : \theta \in \Theta\}$$

of closed sets in Z, and

$$h : \mathcal{K}_\rho \to \mathcal{K}_\rho,$$

$$\vartheta : \mathcal{K}_\rho \to \Theta$$

satisfying the following conditions:

(a) *for each K in $\mathcal{K}_\rho$, the set*

$$h(K) = K \cap S(\vartheta(K))$$

is nonempty and of ρ-diameter less than ϵ;

(b) *if $K_1 \subset K_2$ and $h(K_2) \cap K_1 \neq \emptyset$, then*

$$\vartheta(K_1) = \vartheta(K_2)$$

and

$$h(K_1) = h(K_2) \cap K_1;$$

(c) *if $h(K_1) \cap h(K_2) \neq \emptyset$, then*

$$\vartheta(K_1) = \vartheta(K_2);$$

(d) *if $\vartheta(K_1) = \vartheta(K_2)$, then*

$$\rho\text{-diam}\,(h(K_1) \cup h(K_2)) < \epsilon;$$

(e) *if A is a directed set and $\{K_\alpha : \alpha \in A\}$ is a decreasing net of nonempty compact sets with*

$$K = \bigcap\{K_\alpha : \alpha \in A\},$$

then for some $\alpha \in A$,

$$\vartheta(K_\beta) = \vartheta(K),$$

for all β beyond α in A;

(f) *if F is any upper semi-continuous set-valued function, from a metric space X to Z, taking only nonempty compact values, the sets*

$$X(\theta) = \{x : \vartheta(F(x)) = \theta\}, \quad \theta \in \Theta,$$

form a disjoint discretely σ-decomposable family of $\mathcal{F}_\sigma$-sets.

Proof. Since Z is σ-fragmented by ρ using closed sets, given $\epsilon > 0$, we can write

$$Z = \bigcup_{n=0}^{\infty} Z_n$$

with $Z_0 = \emptyset$, and for $n \geq 1$, Z_n a closed set with the property that each non-empty subset contains a nonempty relatively open subset of diameter less than ϵ. Except in the trivial case when Z is finite we may suppose that $Z_n \backslash Z_{n-1}$ is nonempty for each $n \geq 1$. For each $n \geq 1$ we apply Lemma 4.2 to the set

$$D_n = Z_n \backslash Z_{n-1}$$

and choose a disjoint family

$$\{B(n, \gamma) : 0 \leq \gamma < \Gamma(n)\}$$

of $\mathcal{F}_\sigma$-sets in Z with the following properties:

(α) the union

$$\bigcup\{B(n, \beta) : 0 \leq \beta \leq \gamma\}$$

is relatively open in D_n for $0 \leq \gamma < \Gamma(n)$;

(β) $B(n, \gamma)$ is nonempty with

$$\rho\text{-diam cl } B(n, \gamma) < \epsilon$$

for $0 \leq \gamma < \Gamma(n)$;

(γ) $\bigcup\{B(n, \gamma) : 0 \leq \gamma < \Gamma(n)\} = D_n$.

We write

$$R(n, \gamma) = D_n \backslash \bigcup\{B(n, \beta) : 0 \leq \beta < \gamma\}.$$

Since

$$\bigcup\{B(n, \beta) : 0 \leq \beta < \gamma\} = \bigcup\left\{\bigcup\{B(n, \alpha) : 0 \leq \alpha \leq \beta\} : 0 \leq \beta < \gamma\right\},$$

the set $R(n, \gamma)$ is relatively closed in $D_n = Z_n \backslash Z_{n-1}$. Since Z_n is closed, the set

$$S(n, \gamma) = \text{cl } R(n, \gamma)$$

is a closed subset of Z_n and

$$R(n, \gamma) = S(n, \gamma) \backslash Z_{n-1}.$$

Note that, since the sets $R(n, \gamma)$ decrease as γ increases, so do the sets $S(n, \gamma)$. For $0 \leq \gamma + 1 < \Gamma(n)$, we write

$$T(n, \gamma) = Z_{n-1} \cup S(n, \gamma + 1).$$

We remark that, eventually, after a change of notation, the family

$$\{S(n, \gamma) : 0 \leq \gamma < \Gamma(n), \ 1 \leq n\}$$

will become the family

$$\{S(\theta) : \theta \in \Theta\}.$$

We now introduce a function χ on $\mathcal{K}_\rho$ with values of the form

$$\chi(K) = (n(K), \gamma(K)),$$

with

$$0 \leq \gamma(K) < \Gamma(n(K)), \quad 1 \leq n(K).$$

We define $n(K)$ to be the least integer n such that

$$K \cap Z_n \neq \emptyset.$$

Then $n(K) \geq 1$ and

$$K \cap Z_{n(K)-1} = \emptyset.$$

For the rest of this paragraph, we give n the fixed value $n(K)$. For $0 \leq \gamma < \Gamma(n)$, we have

$$K \cap \left(D_n \setminus \bigcup \{B(n, \beta) : 0 \leq \beta < \gamma\} \right) = K \cap R(n, \gamma)$$

$$= K \cap (S(n, \gamma) \setminus Z_{n-1})$$

$$= K \cap S(n, \gamma),$$

since $K \cap Z_{n-1} = \emptyset$. Thus

$$K \cap S(n, \gamma), \ 0 \leq \gamma < \Gamma(n),$$

is a decreasing sequence of compact sets and

$$\bigcap \{K \cap S(n, \gamma) : 0 \leq \gamma < \Gamma(n)\} = K \cap (D_n \setminus \{B(n, \gamma) : 0 \leq \gamma < \Gamma(n)\})$$

$$= \emptyset.$$

Hence, for some γ, $0 \leq \gamma < \Gamma(n)$, we have

$$K \cap S(n, \gamma) = \emptyset.$$

There will be a least ordinal γ with this property. If γ were a limit ordinal, then the sequence

$$K \cap S(n, \beta), \quad 0 \leq \beta < \gamma,$$

would be a decreasing sequence of nonempty compact sets with empty intersection. Thus there is an ordinal $\gamma(K)$, say, with $1 \leq \gamma(K) + 1 < \Gamma$,

$$K \cap S(n(K), \gamma(K)) \neq \emptyset,$$

$$K \cap S(n(K), \gamma(K) + 1) = \emptyset.$$

We define χ so that $\chi(K)$ has the value

$$\chi(K) = (n(K), \gamma(K)).$$

Note that we have chosen $n(K)$, $\gamma(K)$ so that

$$K \cap Z_{n(K)-1} = \emptyset,$$

$$K \cap S(n(K), \gamma(K) + 1) = \emptyset,$$

$$K \cap S(n(K), \gamma(K)) \neq \emptyset.$$

Writing

$$T(n, \gamma) = Z_{n-1} \cup S(n, \gamma + 1)$$

we note that $n(K)$, $\gamma(K)$ satisfy

$$K \cap T(n(K), \gamma(K)) = \emptyset,$$

$$K \cap S(n(K), \gamma(K)) \neq \emptyset.$$

Now, if $1 \leq n^*$, $1 \leq \gamma^* < \Gamma(n^*)$, and

$$K \cap T(n^*, \gamma^*) = \emptyset,$$

$$K \cap S(n^*, \gamma^*) \neq \emptyset,$$

we show that we must have $n^* = n(K)$, $\gamma^* = \gamma(K)$. Since

$$T(n^*, \gamma^*) \supset Z_{n^*-1},$$

$$S(n^*, \gamma^*) \subset Z_{n^*},$$

we have

$$K \cap Z_{n^*-1} = \emptyset, \quad K \cap Z_{n^*} \neq \emptyset,$$

ensuring that $n^* = n(K)$. Then

$$K \cap S(n^*, \gamma^* + 1) = \emptyset, \quad K \cap S(n^*, \gamma^*) \neq \emptyset,$$

ensuring that $\gamma^* = \gamma(K)$.

We now define a function h from $\mathcal{K}_\rho$ to $\mathcal{K}_\rho$ by taking

$$h(K) = K \cap S(n(K), \gamma(K)).$$

(We regard h both as a point-valued function from the space $\mathcal{K}_\rho$ to the space $\mathcal{K}_\rho$ and also as a set-valued function from $\mathcal{K}_\rho$ to Z.) By the choice of $n(K)$, $\gamma(K)$, the set $h(K)$ is always a nonempty compact set in Z. Further, since

$$K \subset Z \backslash Z_{n(K)-1},$$

$$K \cap S(n(K), \gamma(K) + 1) = \emptyset,$$

we have

$$h(K) = K \cap S(n(K), \gamma(K))$$

$$= K \cap \left((S(n(K), \gamma(K)) \backslash Z_{n(K)-1}) \backslash (S(n(K), \gamma(K) + 1) \backslash Z_{n(K)-1}) \right)$$

$$= K \cap \left(R(n(K), \gamma(K)) \backslash R(n(K), \gamma(K) + 1) \right)$$

$$= K \cap B(n(K), \gamma(K)).$$

Since $S(n(K), \gamma(K))$ is closed and $B(n(K), \gamma(K))$ is of ρ-diameter less than ϵ, it follows that $h(K)$ is a nonempty compact subset of K of ρ-diameter less than ϵ.

Now suppose that $K_1 \subset K_2$ and that $h(K_2) \cap K_1 \neq \emptyset$. Since $K_1 \subset K_2$, we have

$$n(K_2) \leq n(K_1).$$

If we had

$$n(K_2) < n(K_1),$$

then we would have

$$h(K_2) \subset Z_{n(K_2)},$$

$$K_1 \cap Z_{n(K_2)} = \emptyset,$$

which would be impossible since $h(K_2) \cap K_1 \neq \emptyset$. Hence

$$n(K_1) = n(K_2).$$

Now, since $K_1 \subset K_2$, we have

$$\gamma(K_1) \leq \gamma(K_2).$$

If we had

$$\gamma(K_1) < \gamma(K_2),$$

then we would have

$$h(K_2) \subset S(n(K_2), \gamma(K_2)),$$

$$K_1 \cap S(n(K_2), \gamma(K_1) + 1) = \emptyset,$$

and we would again have a contradiction. Thus we must have

$$\gamma(K_1) = \gamma(K_2).$$

Now

$$h(K_1) = K_1 \cap S(n(K_1), \gamma(K_1)),$$

$$h(K_2) = K_2 \cap S(n(K_1), \gamma(K_1))$$

and $K_1 \subset K_2$ implies that

$$h(K_1) = h(K_2) \cap K_1.$$

Further, if

$$h(K_1) \cap h(K_2) \neq \emptyset,$$

then

$$B(n(K_1), \gamma(K_1)) \cap B(n(K_2), \gamma(K_2)) \neq \emptyset,$$

so that

$$(n(K_1), \gamma(K_1)) = (n(K_2), \gamma(K_2)).$$

Finally, if

$$(n(K_1), \gamma(K_1)) = (n(K_2), \gamma(K_2)),$$

then

$$h(K_1) \cup h(K_2)$$

is contained in

$$B(n(K_1), \gamma(K_2)),$$

and so has ρ-diameter less than ϵ.

Now let A be any directed set and consider a decreasing net $\{K_\alpha : \alpha \in A\}$ of nonempty compact sets in Z with intersection

$$K = \bigcap \{K_\alpha : \alpha \in A\}.$$

Since

$$K \cap S(n(K), \gamma(K)) \neq \emptyset,$$

we have

$$K_\alpha \cap S(n(K), \gamma(K)) \neq \emptyset$$

80   CHAPTER 4

for each α in A. In particular,

$$K_\alpha \cap Z_{n(K)} = \emptyset$$

for each α in A. We also have

$$K \cap Z_{n(K)-1} = \emptyset.$$

Since $\{K_\alpha \cap Z_{n(K)-1} : \alpha \in A\}$ is a decreasing net of compact sets with intersection

$$K \cap Z_{n(K)-1} = \emptyset,$$

there must be some α in A with

$$K_\alpha \cap Z_{n(K)-1} = \emptyset.$$

Hence

$$K_\beta \cap Z_{n(K)-1} = \emptyset$$

and

$$n(K_\beta) = n(K)$$

for all β beyond α in A. Since

$$K \cap T(n(K), \gamma(K)) = \emptyset,$$

we have

$$K \cap S(n(K), \gamma(K) + 1) = \emptyset.$$

As before, by considering the directed subfamily

$$\{K_\beta \cap S(n(K), \gamma(K) + 1) : \beta > \alpha, \ \beta \in A\},$$

we can choose β beyond α, so that

$$K_\delta \cap S(n(K), \gamma(K) + 1) = \emptyset$$

and

$$\gamma(K_\delta) = \gamma(K)$$

for all δ beyond β in A. Thus

$$\chi(K_\delta) = \chi(K) = (n(K), \gamma(K))$$

for all δ beyond β in A.

Now consider any upper semi-continuous set-valued function F, from a metric space X to Z, taking only nonempty compact values. We define a family of sets

$$\{X(n, \gamma) : 1 \le n, \ 0 \le \gamma < \Gamma(n)\}$$

in X by taking

$$X(n, \gamma) = \{x : \chi(F(x)) = (n, \gamma)\}.$$

These sets are all disjoint. We consider separately the countable sequence of families

$$\{X(n, \gamma) : 0 \leq \gamma < \Gamma(n)\}, \quad n \geq 1.$$

For fixed n and $0 \leq \gamma < \Gamma(x)$, we have

$X(n, \gamma)$

$$= \{x : F(x) \cap S(n, \gamma) \neq \emptyset \text{ and } F(x) \cap T(n, \gamma) = \emptyset\}$$

$$= \{x : F(x) \cap S(n, \gamma) \neq \emptyset, \ F(x) \cap Z_{n-1} = \emptyset \text{ and } F(x) \cap S(n, \gamma + 1) = \emptyset\}$$

$$= \{x : F(x) \cap Z_{n-1} = \emptyset\} \cap (G(n, \gamma + 1) \backslash G(n, \gamma)),$$

where we write

$$G(n, \beta) = \{x : F(x) \cap S(n, \beta) = \emptyset\}$$

for $0 \leq \beta < \Gamma$. Since $S(n, \beta)$ is closed and F is upper semi-continuous, each set

$$G(n, \beta), \quad 0 \leq \beta < \Gamma,$$

is an open set in X. When β is a successor ordinal, write

$$Y(n, \beta) = G(n, \beta) \backslash G(n, \beta - 1)$$

$$= G(n, \beta) \backslash \bigcup \{G(n, \delta) : 0 \leq \delta < \beta\}.$$

When β is a limit ordinal write

$$Y(n, \beta) = G(n, \beta) \backslash \bigcup \{G(n, \delta) : 0 \leq \delta < \beta\}$$

$$= \{x : F(x) \cap S(n, \beta) = \emptyset\} \backslash \bigcup \{\{x : F(x) \cap S(n, \delta) = \emptyset\} : 0 \leq \delta < \beta\}.$$

Since $F(x) \cap S(n, \delta), 0 \leq \delta < \beta$, is a decreasing sequence of compact sets,

$$F(x) \cap S(n, \beta) = \emptyset$$

implies that

$$F(x) \cap S(n, \delta) = \emptyset$$

for some $\delta, 0 \leq \delta < \beta$. Hence $Y(n, \beta) \neq \emptyset$ when β is a limit ordinal. Now the family

$$\{Y(n, \beta) : 0 \leq \beta < \Gamma(n)\}$$

differs from the family

$$\{G(n, \gamma + 1)\backslash G(n, \gamma) : 0 \leq \gamma < \Gamma\}$$

only by the inclusion of certain empty sets (those corresponding to limit ordinals β). Since $Y(n, \beta)$ has the form

$$Y(n, \beta) = G(n, \beta)\backslash \bigcup\{G(n, \delta) : 0 \leq \delta < \beta\}$$

with the sets $G(n, \beta)$ open in the metric space X, it follows, by Lemma 2.6, that the family

$$\{Y(n, \beta) : 0 \leq \beta < \Gamma(n)\},$$

and so also the family

$$\{G(n, \gamma + 1)\backslash G(n, \gamma) : 0 \leq \gamma < \Gamma\},$$

are discretely σ-decomposable families of $\mathcal{F}_\sigma$-sets. Since

$$\{x : F(x) \cap Z_{n-1} = \emptyset\}$$

is open, it follows that the whole family

$$X(n, \gamma) = \{x : F(x) \cap Z_{n-1} = \emptyset\} \cap (G(n, \gamma + 1)\backslash G(n, \gamma)),$$

$0 \leq \gamma < \Gamma(n)$, $1 \leq n$ is a disjoint discretely σ-decomposable family of $\mathcal{F}_\sigma$-sets.

It now remains to define the families and functions that we require in terms of the families and functions that we have constructed. We take

$$\Theta = \{\theta = (n, \gamma) : 0 \leq \gamma < \Gamma(n), \ 1 \leq n\}$$

and define

$$S(\theta) = S(n, \gamma), \quad \text{with } (n, \gamma) = \theta.$$

The function $h : \mathcal{K}_\rho \rightarrow \mathcal{K}_\rho$ remains unchanged. We take

$$\vartheta(K) = (n(K), \gamma(K)).$$

It is easy to check that our requirements are satisfied. $\square$

Proof of Theorem 4.5. For $\epsilon = 2^{-n}$ and $n = 1, 2, \ldots,$ we use Lemma 4.3 to construct an index set $\Theta(n)$, a family

$$\{S_n(\theta) : \theta \in \Theta(n)\}$$

of closed sets in Z, and functions

$$h_n : \mathcal{K}_\rho \rightarrow \mathcal{K}_\rho,$$

$$\vartheta_n : \mathcal{K}_\rho \to \Theta(n)$$

satisfying the following conditions:

(a) for each K in $\mathcal{K}_\rho$, the set

$$h_n(K) = K \cap S_n(\vartheta_n(K))$$

is nonempty and of ρ-diameter less than 2^{-n};

(b) if $K_1 \subset K_2$ and $h_n(K_2) \cap K_1 \neq \emptyset$, then

$$\vartheta_n(K_1) = \vartheta_n(K_2)$$

and

$$h_n(K_1) = h_n(K_2) \cap K_1;$$

(c) if $h_n(K_1) \cap h_n(K_2) \neq \emptyset$, then

$$\vartheta_n(K_1) = \vartheta_n(K_2);$$

(d) if $\vartheta_n(K_1) = \vartheta_n(K_2)$, then

$$\rho\text{-diam}\,(h_n(K_1) \cup h_n(K_2)) < 2^{-n};$$

(e) if A is a directed set and $\{K_\alpha : \alpha \in A\}$ is a decreasing net of nonempty compact sets with

$$K = \bigcap \{K_\alpha : \alpha \in A\},$$

then for some $\alpha \in A$,

$$\vartheta_n(K_\beta) = \vartheta_n(K)$$

for all β beyond α in A;

(f) if F is any upper semi-continuous set-valued function, from a metric space X to Z, taking only nonempty compact values, the sets

$$X_n(\theta) = \{x : \vartheta_n(F(x)) = \theta\}, \quad \theta \in \Theta(n),$$

form a disjoint discretely σ-decomposable family of $\mathcal{F}_\sigma$-sets.

It is convenient to write $h_0(K) = k_0(K) = K$, and to write $\Phi(0) = \{\emptyset\}$ for the set of empty sequences of length zero. For $n \geq 1$, let $\Phi(n)$ denote the index set consisting of all sequences

$$\overline{\omega}_n(K) = \varphi_1, \varphi_2, \ldots, \varphi_n$$

defined by

$$\varphi_i = \vartheta_i(k_{i-1}(K)), \quad 1 \leq i \leq n,$$

with

$$k_i(K) = h_i(k_{i-1}(K)), \quad 1 \le i \le n,$$

for some K in $\mathcal{K}_\rho$. Write

$$T_n(\overline{\omega}_n(K)) = S_1(\varphi_1) \cap S_2(\varphi_2) \cap \cdots \cap S_n(\varphi_n).$$

We verify that

$$k_n(K) = K \cap T_n(\overline{\omega}_n(K)).$$

To see this, note that

$$k_n(K) = h_n(k_{n-1}(K))$$

$$= k_{n-1}(K) \cap S_n(\vartheta_n(k_{n-1}(K)))$$

$$= k_{n-1}(K) \cap S_n(\varphi_n).$$

Applying this inductively

$$k_n(K) = K \cap S_1(\varphi_1) \cap S_2(\varphi_2) \cap \cdots \cap S_n(\varphi_n)$$

$$= K \cap T_n(\overline{\omega}_n(K)),$$

since $k_0(K) = K$. Note that we defined functions

$$\overline{\omega}_n : \mathcal{K}_\rho \to \Phi(n)$$

and

$$k_n : \mathcal{K}_\rho \to \mathcal{K}_\rho,$$

with

$$k_n(K) = h_n \circ h_{n-1} \circ \cdots \circ h_2 \circ h_1(K)$$

$$= K \cap T_n(\overline{\omega}_n(K)).$$

Now define the function

$$s : \mathcal{K}_\rho \to \mathcal{K}_\rho$$

by

$$s(K) = \bigcap_{n=1}^{\infty} k_n(K).$$

We verify that $s(K)$ is a single point in Z, and that the corresponding function $s : \mathcal{K}_\rho \to Z$ satisfies our requirements.

The condition (a) ensures that the sequence

$$K, k_1(K), k_2(K), \ldots$$

is a decreasing sequence of nonempty compact sets with ρ-diameter tending to zero. Hence $s(K)$ is a well-defined singleton contained in K, for each K in $\mathcal{K}_\rho$. Thus (a) of Theorem 4.5 holds.

Suppose that K_1, K_2 are two sets of $\mathcal{K}_\rho$ with

$$s(K_2) \in K_1 \subset K_2.$$

Then $s(K_2) \in h_1(K_2)$ so that $K_1 \subset K_2$ and $h_1(K_2) \cap K_1 \neq \emptyset$. By condition (b) we have

$$\vartheta_1(K_1) = \vartheta_1(K_2) \text{ and } h_1(K_1) = h_1(K_2) \cap K_1.$$

Thus

$$s(K_2) \in h_1(K_1) \subset h_1(K_2)$$

or

$$s(K_2) \in k_1(K_1) \subset k_1(K_2).$$

Applying this argument inductively, we find that

$$s(K_2) \in k_n(K_1) \subset k_n(K_2)$$

for all $n \geq 1$. Hence

$$s(K_2) \in \bigcap_{n=1}^{\infty} k_n(K_1) = \{s(K_1)\}.$$

Thus

$$s(K_1) = s(K_2)$$

and (b) of Theorem 4.5 holds.

Now suppose that K_1, K_2 are sets of $\mathcal{K}_\rho$ with

$$\overline{\omega}_n(K_1) = \overline{\omega}_n(K_2) = \varphi_1, \varphi_2, ..., \varphi_n.$$

Then

$$k_n(K_i) = h_n\big(k_{n-1}(K_i)\big)$$

for $i = 1, 2$ and

$$\vartheta_n\big(k_{n-1}(K_1)\big) = \vartheta_n\big(k_{n-1}(K_2)\big) = \varphi_n.$$

By the condition (d),

$$\rho\text{-diam}\big(k_n(K_1) \cup k_n(K_2)\big) < 2^{-n}.$$

Let A be any directed set and let $\{K_\alpha : \alpha \in A\}$ be a decreasing net of nonempty compact sets in Z with

$$K = \bigcap\{K_\alpha : \alpha \in A\}.$$

Let $n \geq 1$ be given. By the condition (e), there is an α_1 in A with

$$\vartheta_1(K_\beta) = \vartheta_1(K)$$

for all β beyond α_1 in A. For all such β, we have

$$k_1(K_\beta) = K_\beta \cap T_1(\vartheta_1(K)),$$

$$k_1(K) = K \cap T_1(\vartheta_1(K)).$$

Applying the same argument to the net

$$\{k_1(K_\beta) : \beta \in A, \ \beta > \alpha_1\} = \left\{K_\beta \cap T_1(\vartheta_1(K)) : \beta \in A, \ \beta > \alpha_1\right\}$$

of nonempty compact sets and the function h_2, we obtain an $\alpha_2 > \alpha_1$ so that

$$\vartheta_2\left(k_1(K_\beta)\right) = \vartheta_2(k_1(K))$$

for all β beyond α_2 in A. After n steps, we obtain α_n in A such that

$$\overline{\omega}(K_\beta) = \overline{\omega}_n(K)$$

for all β beyond α_n in A. This ensures that

$$\rho\text{-diam}\left(k_n(K_\beta) \cup k_n(K)\right) < 2^{-n}$$

for all β beyond α_n in A. Thus

$$\rho\left(s(K_\beta), s(K)\right) < 2^{-n}$$

for all β beyond α_n in A. This shows that $s(K_\beta)$ converges to $s(K)$ in Z with the metric ρ, following the directed set A. So conclusion (c) of Theorem 4.5 follows.

Let F be any upper semi-continuous set-valued function, from a metric space X to Z, taking only nonempty compact values. We show that, for each $n \geq 1$, the sets

$$\Xi_n(\varphi) = \{x : \overline{\omega}_n(F(x)) = \varphi\}, \quad \varphi \in \Phi(n),$$

form a disjoint discretely σ-decomposable family of $\mathcal{F}_\sigma$-sets in X. By the result (f), this is true when $n = 1$. Suppose that it is true for some $n \geq 1$. Consider any

$$\varphi = \varphi_1, \varphi_2, ..., \varphi_n, \varphi_{n+1}$$

in $\Phi(n+1)$. Then

$$\Xi_{n+1}(\varphi) = \{x : \overline{\omega}_{n+1}(F(x)) = \varphi\}$$

$$= \{x : \overline{\omega}_n(F(x)) = \varphi \mid n\} \cap \{x : \vartheta_{n+1}(k_n(F(x))) = \varphi_{n+1}\}$$

$$= \Xi_n(\varphi \mid n) \cap \{x : \vartheta_{n+1}(F(x) \cap T_n(\varphi \mid n)) = \varphi_{n+1}\}.$$

On the set $\Xi_n(\varphi \mid n)$ the set-valued function

$$F(x) \cap T_n(\varphi \mid n)$$

is upper semi-continuous and takes only nonempty compact values in Z. It follows by condition (f) that, for each fixed $\chi \in \Phi(n)$, the family

$$\{\Xi_{n+1}(\varphi) : \varphi \in \Phi(n+1), \; \varphi \mid n = \chi\}$$

is a disjoint discretely σ-decomposable family of $\mathcal{F}_\sigma$-sets in the space $\Xi_n(\chi)$, considered relative to $\Xi_n(\chi)$. By Lemma 2.3, it follows that the family

$$\{\Xi_{n+1}(\varphi) : \varphi \in \Phi(n+1)\}$$

is a disjoint discretely σ-decomposable family of $\mathcal{F}_\sigma$-sets. We now show that the family

$$\{\Xi_n(\varphi) : \varphi \in \Phi(n), \; n \geq 1\}$$

is a discretely σ-decomposable family of $\mathcal{F}_\sigma$-sets that forms a base for the function

$$f = s \circ F : X \to (Z, \rho).$$

Let G be any ρ-open set in Z; we show that $f^{-1}(G)$ is the union of the sets $\Xi_n(\varphi)$ that it contains. Consider any point x^* in $f^{-1}(G)$. Then $f(x^*) = s \circ F(x^*) \in G$. Choose n so large that the spherical ball with center $f(x^*)$ and radius 2^{-n} is contained in G. Write $\varphi = \overline{\omega}(F(x^*))$. Then

$$s \circ F(x^*) \subset k_n(F(x^*)).$$

If x is any point of $\Xi_n(\varphi)$, then $\overline{\omega}(F(x)) = \varphi$. By the result (d)

$$\rho\text{-diam}\,(k_n(F(x)) \cup k_n(F(x^*))) < 2^{-n}.$$

Since

$$f(x^*) = s \circ F(x^*) \subset k_n(F(x^*)),$$

we have

$$k_n(F(x)) \subset G$$

and

$$f(x) = s \circ F(x) \in k_n(F(x)) \subset G.$$

Thus

$$x \in f^{-1}(G).$$

This shows that

$$x^* \in \Xi_n(\varphi) \subset f^{-1}(G).$$

So $f^{-1}(G)$ is the union of the sets $\Xi_n(\varphi)$, $\varphi \in \Phi(n)$, $n \geq 1$, that it contains.

It now follows that f is a σ-discrete function of the first Borel class.

If ρ has the ϵ-neighborhood property, it follows from the definition of this property, that the representation function

$$R : \mathcal{K}_\rho \to Z,$$

given by

$$R(K) = K,$$

is an upper semi-continuous set-valued function, from $\mathcal{K}_\rho$ to Z, taking only nonempty compact values. By the result of the last paragraph, the selector

$$s = s \circ R$$

is a σ-discrete function of the first Borel class from $\mathcal{K}_\rho$ to (Z, ρ). $\square$

Proof of Theorem 4.2. The result follows immediately from Theorem 4.5 since the metric ρ has the ϵ-neighborhood property for the compact sets of (Y, ρ) and (Y, ρ) is automatically fragmented by ρ. $\square$

Proof of Theorems 4.3 and 4.4. The results follow immediately from Theorem 4.5 on using Lemma 4.1 and Theorem 2.1.

4.3 REMARKS

1. As we have already remarked, Theorem 4.4 is closely related to a result of Ghoussoub, Maurey and Schachermayer [14]. Their result is less general than Theorem 4.4 in that they need to assume that their space is fragmented rather than σ-fragmented by closed sets; it is more general in that they are able to work with their concept of "a slice-upper semi continuous multivalued map". They also obtain selections from closed bounded sets into certain types of their extreme points.

2. Ghoussoub, Maurey and Schachermayer [14, p. 489] state that a function between two topological spaces X and Y is Baire-1 if the inverse image of every open subset of Y is an $\mathcal{F}_\sigma$-set in X. Such a function would more usually be called a function of the first Borel class; functions of the first Baire class are usually defined as functions that are pointwise limits of continuous functions. The two classes do not coincide even when X and Y are both metric spaces (see Example 2.1 above). This means that their paper needs to be read with this distinction in mind. However, the two concepts are equivalent when X is a metric space and Y is a convex subset of normed linear space.

3. Many metric spaces can be expressed as the continuous image of the positive real axis $[0, \infty)$. It is easy to verify that the proof of Theorem 4.1 applies with minor modifications to the closed sets of any such space.

4. In the proof of the key Lemma 4.3 we make a partial selection by combining the method of first choice, in fixing $n(K)$, and the method of deferred choice, in fixing $\gamma(K)$. It might be possible, and would probably be more elegant, if the proof could be more closely modeled on the proof of Theorem 4.1.

5. If in the proof of Theorem 4.1, the curve $p : [0, 1] \to Q$ is taken to be the rather special Peano curve constructed relatively recently by Pach and Rogers [65], it is easy to verify that the selector s for $\mathcal{K}$ satisfies the following mysterious extra condition:

(d) if K is any closed set of $\mathcal{K}$, the set

$$K^* = \bigcup \{s(H) : H \in \mathcal{K} \text{ and } H \supset K\}$$

is a closed convex subset of Q that meets K just at the point $s(K)$.

6. While we have not explicitly written the statements into the conclusions, all the selectors in Theorems 4.2–4.5, being σ-discrete and of the first Borel class, will be such that their points of discontinuity are $\mathcal{F}_\sigma$-sets of the first Baire category in X, just as in the theorems of Chapter 3.

Chapter 5

Applications

In this chapter we mainly use theorems that we have already established to give results for a variety of set-valued maps that have geometric origins. In particular we study subdifferentials, two sorts of attainment maps, metric projections and continuous maps to a space of compact convex sets. For these purposes we study maximal monotone maps. We also give results that act as partial converses to Theorems 3.2 and 3.3.

A set-valued map F from a Banach space X to its dual space X^* is said to be a *monotone map* if

$$\langle x_2 - x_1, x_2^* - x_1^* \rangle \geq 0$$

for all choices of x_1, x_1^*, x_2, x_2^* with

$$x_1^* \in F(x_1) \text{ and } x_2^* \in F(x_2).$$

Here we allow F to take the value $\emptyset$; we use

$$D(F) = \{x : F(x) \neq \emptyset\}$$

to denote the *domain* of F. A set-valued map F from X to X^* is said to be a *maximal monotone map*, if it is a monotone map and it has a graph

$$Gr(F) = \bigcup \{\{x\} \times F(x) : x \in X\}$$

that is a proper subset of the graph of no monotone map from X to X^*.

Suppose that

$$F(x) = \bigcup \{F_\gamma(x) : 0 \leq \gamma < \Gamma\}$$

for all x in X, with $\{F_\gamma : 0 \leq \gamma < \Gamma\}$ a transfinite sequence of monotone maps with increasing graphs. If

$$x_1^* \in F(x_1) \text{ and } x_2^* \in F(x_2),$$

then for some γ, $0 \leq \gamma < \Gamma$, we have

$$x_1^* \in F_\gamma(x_1) \text{ and } x_2^* \in F_\gamma(x_2),$$

so that

$$\langle x_2 - x_1, x_2^* - x_2 \rangle \geq 0.$$

Thus F is necessarily a monotone map whose graph includes the graphs of all the maps $F_\gamma, 0 \leq \gamma < \Gamma$. It follows by Zorn's lemma that *every monotone map from X to X^* has a graph that is contained in the graph of some maximal monotone map from X to X^**.

In Section 5.1 we develop some of the known properties of maximal monotone maps (see Theorem 5.9) and prove the following selection result.

Theorem 5.1 *Let X be a Banach space with dual space X^*. Suppose that each weak* compact set in X^* is weak* fragmented. Let H be a maximal monotone map from X to X^*. Let D be the domain of H and let D_0 be the interior of D. Let U_0 be the set of points of D_0 at which H is point-valued and norm upper semi-continuous.*

Then U_0 is a dense G_δ-subset of D_0.

The map H has a selector h from D to $(X^*, norm)$ that is of the first Baire class.

For each selector f for H from D_0 to $(X^*, norm)$, of any kind, the set of points where f is norm continuous coincides with U_0.

It is convenient to introduce the extended real line $\mathbb{R}_\infty = \mathbb{R} \cup \{+\infty\}$, topologized by taking the intervals of the form $(t, +\infty]$ as a base for the neighborhoods of the point $+\infty$. When X is a Banach space and f is a continuous convex function from X to $\mathbb{R}_\infty$, the subdifferential $D_x f$ of f at a point x of X is defined to be the set of elements x^* of X^* satisfying the condition

$$f(x) + \langle u, x^* \rangle \leq f(x + u)$$

for all u in X. Except when f only takes the value $+\infty$, this subdifferential D_f turns out to be a maximal monotone map. We verify this in Section 5.2 and then prove the following theorem.

Theorem 5.2 *Let X be a Banach space with dual space X^*.*

If each weak* compact subset of X^* is weak* fragmented, then the subdifferential map D_f corresponding to each continuous convex function from X to $\mathbb{R}_\infty$, that is not identically $+\infty$, has a selector of the first norm Baire class on the domain D of D_f and so is continuous at all points of a G_δ-set dense in D. If the subdifferential map of each continuous convex function from X to $\mathbb{R}$ has a selector that is norm continuous at all points of a G_δ-set dense in X, then X is an Asplund space.

Since the subdifferential map D_f of a continuous convex function from X to $\mathbb{R}$ is necessarily a weak* upper semi-continuous set-valued map taking only nonempty weak* compact convex values, the second part of this theorem, taken with the fact that the weak* compact subsets of the dual of an Asplund space are weak* fragmented, provides a strong converse to Theorem 3.3.

Let K be a nonempty weakly compact set in a Banach space X with dual space X^*. The *attainment map from X^* to K* is defined to be the set-valued map $F_K : X^* \to K$, with

$$F_K(x^*) = \{x \in K : \langle x, x^* \rangle = \sup\{\langle k, x^* \rangle : k \in K\}\}$$

for all x^* in X^*. In Section 5.3, we discuss such maps and obtain the following theorem.

Theorem 5.3 *Let K be a nonempty weakly compact set in a Banach space X with dual space X^*. Let $F_K : X^* \to K$ be the attainment map for K. Let U^* be the set of points of X^* at which F_K is point-valued and norm upper semi-continuous. Then U^* is a dense G_δ-set in X^*.*

The map F_K has a selector f of the first norm Baire class. Further, for each selector g for F_K the set of points where g is norm continuous coincides with U^*.

Let K be a nonempty weak* compact set in the dual X^* of a Banach space X. *The attainment map from X to K* is defined to be the set-valued map $F_K : X \to K$, with

$$F_K(x) = \{x^* \in K : \langle x, x^* \rangle = \sup\{\langle x, k^* \rangle : k^* \in K\}\}$$

for all x in X. We study such maps in Section 5.4, and we prove the following theorem (see Jayne, Orihuela , Pallarés and Vera [41, Proposition 25, p. 268]).

Theorem 5.4 *Let K be a nonempty weak* compact set in the dual X^* of a Banach space X. Let $F_K : X \to K$ be the attainment map from X to K. Then F_K is a weak* upper semi-continuous map from X to K taking only nonempty weak* compact values.*

If K is convex and weak* fragmented by the norm on X^*, then F_K has a selector $f : X \to (K, norm)$ that is of the first Baire class.

In the case when K is the unit ball of X^*, this selector has been called the Jayne–Rogers selector and has been generalized by Deville, Godefroy and Zizler [7, p. 18 et seq.].

In Chapter 7 we discuss in some detail a partial converse to Theorem 5.4.

Let K be a nonempty set in a Banach space X. For each x in X, we write

$$\rho(x) = \inf\{\|x - k\| : k \in K\}$$

and

$$F(x) = \{k : \|k - x\| = \rho(x) \text{ and } k \in K\}.$$

The set-valued map F is called the nearest point map of K or the metric projection onto K. This map has been much studied, see Kenderov [45] and references given there. We prove two simple results in Section 5.5.

Theorem 5.5 *Let K be a nonempty compact set in a Banach space X. Then the nearest point map F from X to K is weakly upper semi-continuous with nonempty weakly compact values, and so has a selector $f : X \longrightarrow (K, norm)$ of the first Baire class. If the norm on X is strictly convex, then F is a minimal weakly upper semi-continuous map with nonempty weakly compact values, and F is point-valued and norm upper semi-continuous at the points of a dense G_δ-subset of X.*

Theorem 5.6 *Let K be a nonempty weak* closed set in a dual Banach space X^* and suppose that K is weak* fragmented. Then the nearest point map F from X^* to K is weak* upper semi-continuous with nonempty weak* compact values, and so has a selector $f : X^* \longrightarrow (K, norm)$ of the first Baire class. If the norm on X^* is strictly convex, then F is a minimal weak* upper semi-continuous map with nonempty weak* compact values, and F is point-valued and norm upper semi-continuous at the points of a dense G_δ-subset of X.*

So far, we have been content to obtain selectors of the first Baire class that are pointwise limits of continuous functions without obtaining any effective control of the continuous functions tending to the selector. When the set-valued map takes as values subsets of a fixed convex set in a Banach space, we can ensure that the continuous functions converging to the selector also take their values in the convex set (see Theorem 1.4 and the last clause of Theorem 2.1). In Section 5.6 we study a more complicated situation. We prove two theorems that can be regarded as parameterized versions of Theorems 5.5 and 5.3.

Theorem 5.7 *Let X be a metric space and let Y be a Banach space that is σ-fragmented. Let H be a set-valued map from X to Y with nonempty weakly compact convex values, and suppose that H is a continuous map from X to the space of nonempty bounded norm closed sets in Y taken with the Hausdorff metric. Let η be a continuous function from X to Y. For each x in X, write*

$$\rho(x) = \inf\{\|h - \eta(x)\| : h \in H(x)\}$$

and

$$F(x) = \{y \in H : \|y - \eta(x)\| = \rho(x)\}.$$

Then F has a selector f that is the pointwise limit of a sequence $\{h_n\}$ of continuous selectors for H.

Theorem 5.8 *Let X be a Banach space and let Y be a Banach space that is* *σ-fragmented with dual space Y^*. Let H be a set-valued map from X to Y with nonempty weakly compact convex values, and suppose that H is a continuous map from X to the space of nonempty bounded norm closed sets in Y taken with the Hausdorff metric. Let η^* be a continuous function from X to Y^*. For each x in X write*

$$F(x) = \{y \in H(x) : \langle y, \eta^*(x) \rangle = \sup\{\langle h, \eta^*(x) \rangle : h \in H(x)\}\}.$$

Then F has a selector f that is the pointwise limit of a sequence $\{h_n\}$ of continuous selectors for H.

These theorems may perhaps have some practical significance. Consider Theorem 5.7 in the light of the following circumstances. Suppose that the operations of a firm depend on the position of a point x in a parameter space X and that the firm has no control of the position x in X. Suppose that for each x in X the operation of the firm can be specified by the choice of a point y in a Banach space Y, but that, for various practical reasons, y has to be chosen from a nonempty convex weakly compact set $H(x)$. Suppose that, for optimality, for a given value of x, the point y needs to be chosen from $H(x)$ to minimize a continuous real-valued function, say $\langle y, y^*(x) \rangle$, with $y^*(x)$ in Y^*. We suppose that the firm seeks a strategy for its operations in the form of a map

$$h : X \to Y,$$

with

$$h(x) \in H(x), \quad \text{for } x \in X,$$

with h continuous, since the firm wishes to avoid disconcerting discontinuities in its operations, and with h in some sense nearly optimal. Under the conditions of Theorem 5.7, possible solutions are provided by the functions h_n, $n = 1, 2, \ldots$. To find a good choice, further investigation would be necessary. Naturally there has to be a trade-off between the approach of the solution h to the "optimal" solution f and the rate of variation of h with x.

5.1 MONOTONE MAPS AND MAXIMAL MONOTONE MAPS

In this section we develop the parts of the theory of monotone maps and of maximal monotone maps that we need. For a more extensive account of this interesting theory, see [58].

We then prove the selection theorem, Theorem 5.1, stated in the introduction of this chapter.

In the main introduction to this chapter, we have already defined the monotone maps and the maximal monotone maps from a Banach space X to its dual space X^*, and have noted that the graph of a monotone map is always included in the graph of a maximal monotone map. We remark that if F is a monotone map from X to X^*, then the values of F corresponding to distinct points x_1, x_2 in X are "almost separated" in the following way. For any x_1^* in $F(x_1)$ and any x_2^* in $F(x_2)$, we have

$$\langle x_2 - x_1, x_2^* - x_1^* \rangle \geq 0,$$

so that

$$\langle x_2 - x_1, x_1^* \rangle \leq \langle x_2 - x_1, x_2^* \rangle$$

and

$$\sup\{\langle x_2 - x_1, x^* \rangle : x^* \in F(x_1)\} \leq \inf\{\langle x_2 - x_1, x^* \rangle : x^* \in F(x_2)\}.$$

Usually one has strict inequality, in this last inequality, for many pairs x_1, x_2.

We say that a monotone map F is locally bounded on an open set G in X if each point g of G has an open neighborhood $N(g)$ with $\|f(N(g))\|$, defined by

$$\|F(N(g))\| = \sup\{\|x^*\| : x^* \in F(N(g))\}$$

finite. Monotone maps that are nowhere locally bounded do exist (see Example 5.1 below), but they are more difficult to handle. We now give a criterion that ensures that a monotone map is bounded on an open set.

Lemma 5.1 *Let F be a monotone map from a Banach space X to its dual space X^*. For $r > 0$, write*

$$F_r(x) = F(x) \cap (rB^*)$$

for all x in X. If the domain of F_r is dense in an open set G, then

$$\|F(G)\| \leq r.$$

Proof. We suppose that there are points ξ, ξ^* satisfying

$$\xi \in G, \quad \xi^* \in F(\xi), \quad \|\xi^*\| > r,$$

and we seek a contradiction. We first choose an integer n so large that

$$\|\xi^*\| > \frac{n+1}{n-1}\, r,$$

and then choose $\epsilon > 0$ so small that

$$\|\xi^*\| > \frac{(n+1)r}{n-1-n\epsilon}.$$

Since $\xi^* \neq 0$, we can choose η in X with

$$\|\eta\| = 1, \quad \langle \eta, \xi^*/\|\xi^*\| \rangle > 1 - \epsilon.$$

Consider the set $C(\delta)$ of all points x with

$$\|x - \xi - n\delta\eta\| < \delta,$$

with $\delta > 0$. Then $C(\delta)$ is an open set contained in the set

$$\|x - \xi\| < (n + 1)\delta.$$

Since $\xi \in G$, we can choose $\delta > 0$ so small that

$$C(\delta) \subset G.$$

Since the domain of F_r is dense in G, we can choose ζ and ζ^* with

$$\zeta \in C(\delta), \quad \zeta^* \in F(\zeta), \quad \|\zeta^*\| \leq r.$$

Then

$$\|\zeta - \xi - n\delta\eta\| < \delta, \quad \|\zeta - \xi\| < (n + 1)\delta.$$

Since F is a monotone map, we have

$$0 \leq \langle \zeta - \xi, \zeta^* - \xi^* \rangle = \langle \zeta - \xi, \zeta^* \rangle - \langle \zeta - \xi - n\delta\eta, \xi^* \rangle - \langle n\delta\eta, \xi^* \rangle$$

$$< (n + 1)\delta\|\zeta^*\| + \delta\|\xi^*\| - n\delta(1 - \epsilon)\|\xi^*\|$$

$$\leq (n - 1 - n\epsilon)\delta\left[\frac{(n + 1)r}{n - 1 - n\epsilon} - \|\xi^*\|\right] < 0$$

by the choice of n and ϵ. This contradiction proves the lemma. $\quad\square$

Corollary 5.1 *If G is an open set contained in the domain of a monotone map F, then*

$$\sup\{\sup\{\|x^*\| : x^* \in F(x)\} : x \in G\} = \sup\{\inf\{\|x^*\| : x^* \in F(x)\} : x \in G\}.$$

To verify this corollary note that, if

$$\sup\{\inf\{\|x^*\| : x^* \in F(x)\} : x \in G\} = r,$$

then, for each $\epsilon > 0$, the domain of $F_{r+\epsilon}(x) = F(x) \cap (r + \epsilon)B^*$ includes G, so that $\|F(G)\| \leq r + \epsilon$ and $\|F(G)\| \leq r$.

The next lemma gives information about the values taken by a maximal monotone map.

Lemma 5.2 *Let H be a maximal monotone map from a Banach space X to X^*. Then $H(x)$ is weak* closed and convex (and perhaps empty) for each x in X.*

Proof. Fix x_0 in the domain of H and suppose that x_1^*, x_2^* are points of $H(x_0)$. Then for each choice of ξ, ξ^* with

$$\xi \in X, \quad \xi^* \in H(\xi),$$

we have

$$\langle \xi - x_0, \xi^* - x_1^* \rangle \geq 0,$$

$$\langle \xi - x_0, \xi^* - x_2^* \rangle \geq 0,$$

so that

$$\langle \xi - x_0, \xi^* - \{1 - \theta)x_1^* + \theta x_1^* \} \rangle \geq 0$$

for $0 \leq \theta \leq 1$. As H is a maximal monotone map, we must have

$$(1 - \theta)x_1^* + \theta x_2^* \in H(x_0)$$

for $0 \leq \theta \leq 1$. Hence $H(x_0)$ is convex.

Now consider any point x_0^* in the weak* closure of $H(x_0)$. For each choice of ξ, ξ^* with

$$\xi \in X, \quad \xi^* \in H(\xi),$$

and each choice of $\epsilon > 0$, the set of points x^* with

$$\langle \xi - x_0, x^* - x_0^* \rangle \geq -\epsilon$$

is a weak* neighborhood of x_0^*, and so contains a point, x_ϵ^* in $H(x_0)$. Hence

$$\langle \xi - x_0, x_\epsilon^* - x_0^* \rangle \geq -\epsilon$$

and as $x_\epsilon^* \in H(x_0)$,

$$\langle \xi - x_0, \xi^* - x_\epsilon^* \rangle \geq 0.$$

Thus

$$\langle \xi - x_0, \xi^* - x_0^* \rangle \geq -\epsilon$$

for $\epsilon > 0$, so that

$$\langle \xi - x_0, \xi^* - x_0^* \rangle \geq 0.$$

Since this holds for all choices of ξ, ξ^* and H is a maximal monotone map, we must have $x_0^* \in H(x_0)$. Hence $H(x_0)$ is weak* closed, as well as convex. $\quad \square$

Theorem 5.9 *Let H be a maximal monotone map from a Banach space X to X^*. For each $r > 0$, write*

$$H_r(x) = H(x) \cap (rB^*)$$

for $x \in X$. Then the domain $D(H_r)$ is closed and H_r takes only nonempty weak* compact convex values and is weak* upper semi-continuous on $D(H_r)$.

If H has a selector h that is norm continuous at an interior point x_0 of $D(H)$, then $H(x_0)$ reduces to a single point and H is norm upper semi-continuous at x_0.

Proof. We prove that the graph

$$G(H_r) = \{(x, x^*) : x \in X \text{ and } x^* \in H_r(x)\}$$

of H_r is closed in $X \times (X^*, \text{weak}^*)$. Let (ξ, ξ^*) be a point in the closure of this graph. Then $\|\xi^*\| \leq r$. Consider any pair (ζ, ζ^*) of points with

$$\zeta \in X, \quad \zeta^* \in H(\zeta).$$

Let $\epsilon > 0$ be given and write

$$\delta = \frac{\frac{1}{2}\epsilon}{\|\zeta^*\| + r}.$$

The set of points (x, x^*) with

$$\|x - \xi\| < \delta, \langle \zeta - \xi, \zeta^* - x^* \rangle < \langle \zeta - \xi, \zeta^* - \xi^* \rangle + \frac{1}{2}\epsilon$$

is a neighborhood of (ξ, ξ^*) in $X \times (X^*, \text{weak}^*)$. So we can choose a point in the neighborhood, say (η, η^*). Then

$$\|\eta - \xi\| < \delta,$$

$$\langle \zeta - \xi, \zeta^* - \eta^* \rangle < \langle \zeta - \xi, \zeta^* - \xi^* \rangle + \frac{1}{2}\epsilon$$

and

$$\eta^* \in H(\eta), \quad \|\eta^*\| \leq r.$$

Further, as H is a monotone map,

$$\langle \zeta - \eta, \zeta^* - \eta^* \rangle \geq 0.$$

Thus

$$\langle \zeta - \xi, \zeta^* - \xi^* \rangle > \langle \zeta - \xi, \zeta^* - \eta^* \rangle - \frac{1}{2}\epsilon$$

$$= \langle \zeta - \eta, \zeta^* - \eta^* \rangle + \langle \eta - \xi, \zeta^* - \eta^* \rangle - \frac{1}{2}\epsilon$$

$$\geq -\|\eta - \xi\| \cdot \|\zeta^* - \eta^*\| - \frac{1}{2}\epsilon$$

$$\geq -\delta\{\|\zeta^*\| + r\} - \frac{1}{2}\epsilon = -\epsilon.$$

Since ϵ may be arbitrarily small, we must have

$$\langle \zeta - \xi, \zeta^* - \xi^* \rangle \geq 0$$

for all choices of ζ, ζ^* with $\zeta^* \in H(\zeta)$. Since H is a maximal monotone map, we must have $\xi^* \in H(\xi)$ and so also $\xi^* \in H_r(\xi)$. Thus the graph $G(H_r)$ is closed in $X \times (X^*, \text{weak}^*)$.

Now $D(H_r)$ is the projection onto X of the closed set $G(H_r)$ through the weak* compact set rB^*. Hence $D(H_r)$ is closed in X.

By Lemma 5.2, each set $H(x)$, $x \in X$, is weak* closed and convex. Since rB^* is weak* compact and convex, so is

$$H_r(x) = H(x) \cap (rB^*)$$

for each x in X.

Now consider any weak* closed set J in X^*. The set $H_r^{-1}(J)$ is the projection on X of the closed set

$$(X \times J) \cap G(H_r)$$

through the weak* compact set rB^* and so is closed in X. Hence H_r is weak* upper semi-continuous on its closed domain $D(H_r)$ and so also on X.

Finally, suppose that H has a selector h that is norm continuous at an interior point x_0 of $D(H)$. Let $\epsilon > 0$ be given. Then we can choose $\delta > 0$ so that all points x with

$$\|x - x_0\| < \delta$$

lie in $D(H)$ and satisfy

$$\|h(x) - h(x_0)\| < \epsilon.$$

Note that the set function K defined by

$$K(x) = H(x) - h(x_0)$$

$$= \{x^* - h(x_0) : x^* \in H(x)\}$$

is a maximal monotone map from X to X^*, with domain $D(H)$. Further, the open ball $B(x_0, \delta)$ is contained in the domain $D(K_\epsilon)$ of the reduced map K_ϵ defined by

$$K_\epsilon(x) = K(x) \cap (\epsilon B^*)$$

for all x in X. By Lemma 5.1,

$$\|K(B(x_0, \delta))\| \le \epsilon,$$

and so

$$\|x^* - h(x_0)\| \le \epsilon$$

for all pairs x, x^* with

$$\|x - x_0\| \le \delta, \quad x^* \in H(x).$$

Since ϵ may be arbitrarily small, this implies that $H(x_0) = \{h(x_0)\}$ and that H is norm upper semi-continuous at x_0. Further, the rate of the semi-convergence of H to $H(x_0)$ is controlled by the rate at which h converges to $h(x_0)$. More precisely, for all sufficiently small $\delta > 0$,

$$\sup\{\|x^* - h(x_0)\| : x^* \in H(x) \text{ and } \|x - x_0\| < \delta\}$$
$$\leq \sup\{\|h(x) - h(x_0)\| : \|x - x_0\| < \delta\}.$$

Proof of Theorem 5.1. Let H be a maximal monotone map from the Banach space X to its dual X^*, all of whose weak* compact sets are weak* fragmented. For each $r > 0$, write

$$H_r(x) = H(x) \cap (rB^*)$$

for all x in X. Then

$$D = D(H) = \bigcup_{r=1}^{\infty} D(H_r),$$

and D is an $\mathcal{F}_\sigma$-set in X, by Theorem 5.9. Further, each map H_r is a weak* upper semi-continuous map from X to X^* taking only nonempty weak* compact convex values on its closed domain $D(H_r)$. By Theorem 3.3, H_r has a selector, say h_r, on $D(H_r)$ that is σ-discrete and of the first Borel class as a map from $D(H_r)$ to (X^*, norm). Now define a function h on D by taking

$$h(x) = h_1(x), \quad \text{if } x \in D(H_1),$$

$$h(x) = h_r, \quad \text{if } x \in D(H_r) \backslash D(H_{r-1}) \text{ and } r \geq 2.$$

Now h is a selector for H on D that is σ-discrete and of the first Borel class. Hence, by Theorem 2.1, h is of the first Baire class on D.

Now assume that the set D_0 of interior points of D is nonempty, and consider the restriction of h to D_0. Since D_0 is a completely metrizable space, the set V_0 where h is norm continuous is a dense $\mathcal{G}_\delta$-subset of D_0. By Theorem 5.9, H is point-valued and norm upper semi-continuous at each point of V_0. Thus $V_0 \subset U_0$. Conversely, at each point of U_0, H is point-valued and norm upper semi-continuous, forcing h to be norm continuous. Thus U_0 coincides with V_0 and is a dense $\mathcal{G}_\delta$-subset of D_0.

If f is any selector for H, of any kind, on D_0, the same argument shows that the set of points of D_0 where f is norm continuous coincides with U_0. $\square$

5.2 SUBDIFFERENTIAL MAPS

In this section we verify that the subdifferential maps of continuous convex functions from a Banach space X to $\mathbb{R}_\infty$, that are not identically infinite, are maximal monotone maps from X to X^*. We then deduce Theorem 5.2 as a consequence of Theorem 5.1.

We first prove the following lemma giving the properties of subdifferential maps that we shall need to use.

Lemma 5.3   *Let f be a continuous convex function, from a Banach space X to $\mathbb{R}_\infty$, that is finite at at least one point of X. Let D be the set of points x of X for which $f(x)$ is finite. Then D is an open convex set in X. The subdifferential map D_f of f is a maximal monotone map from X to X^* with D as its domain of definition.*

Proof. Since f is convex the set D of points x with $f(x) < +\infty$ is convex. Since f is continuous to $\mathbb{R}_\infty$, the set D is open.

Fix a point a in X with $f(a) < +\infty$. It is clear that if $\xi \notin D$, then $f(\xi) = +\infty$, and we cannot find any ξ^* in X^* satisfying

$$f(\xi) + \langle a - \xi, \xi^* \rangle \le f(a).$$

Thus ξ is not in the domain of D_f. However, if $\xi \in D$, then $f(x)$ is less than $f(\xi) + 1$ for all x in an open set, G say, contained in D. Now the epigraph, A say, of f in $X \times \mathbb{R}$ contains the nonempty open set $G \times (f(x) + 1, +\infty)$. Now $(x, f(x) - \epsilon)$ does not belong to A when $\epsilon > 0$. Thus $(x, f(x))$ is a point on the boundary of the convex set A in $X \times \mathbb{R}$. Now $X \times \mathbb{R}$ is a Banach space. By the Hahn–Banach theorem there is a linear functional, say ℓ, on $X \times \mathbb{R}$, such that

$$\ell(\xi, t) \le \ell(x, f(x))$$

for all points (ξ, t) in A. Now the general linear functional on $X \times \mathbb{R}$ takes the form

$$\langle (x, t), (x^*, t^*) \rangle = \langle x, x^* \rangle + tt^*,$$

with (x^*, t^*) in $X^* \times \mathbb{R}$. Thus, for some ξ^* in X^* and some $\tau^* \in \mathbb{R}$, we have

$$\langle y, \xi^* \rangle + t\tau^* \le \langle x, \xi^* \rangle + f(x)\tau^*$$

for (y, t) in A. In particular, the points $(x, f(x) + p), p \ge 0$, belong to A so that

$$p\tau^* \le 0.$$

Hence $\tau^* \le 0$. The possibility that $\tau^* = 0$ is excluded, since we cannot have

$$\langle y, \xi^* \rangle \le \langle x, \xi^* \rangle$$

for all y in D because x is an interior point of D. Thus τ^* must be negative, and dividing our inequality by $-\tau^*$ and taking $t = f(y)$, we have

$$\langle y, \xi^*/(-\tau^*) \rangle - f(y) \le \langle x, \xi^*/(-\tau^*) \rangle - f(x)$$

or

$$f(x) + \langle y - x, \xi^*/(-\tau^*) \rangle \le f(y)$$

for all y in D. This also holds when $y \notin D$ and $f(y) = +\infty$. Thus $\xi^*/(-\tau^*)$ belongs to $D_x f$, and x belongs to the domain of D_f. Hence the domain of D_f coincides with D.

Now, if ξ, ζ belong to D and

$$\xi^* \in D_\xi f, \quad \zeta^* \in D_\zeta f,$$

we have

$$f(\xi) + \langle \zeta - \xi, \xi^* \rangle \le f(\zeta)$$

and

$$f(\zeta) + \langle \xi - \zeta, \zeta^* \rangle \le f(\xi),$$

so that

$$\langle \zeta - \xi, \zeta^* - \xi^* \rangle = -\langle \zeta - \xi, \xi^* \rangle - \langle \xi - \zeta, \zeta^* \rangle \ge 0.$$

Thus D_f is a monotone map from X to X^*.

Following Minty [58, Theorem 2], we show that D_f is a maximal monotone map from X to X^*. We suppose that $\xi \in X$ and that $\xi^* \notin D_\xi f$, and seek to prove that (ξ, ξ^*) can lie in the graph of no monotone map whose graph contains that of D_f. We need to consider two cases.

Case (a) Suppose that $\xi \in D$. Since $\xi^* \notin D_\xi f$, there is some u in X with

$$f(\xi) + \langle u, \xi^* \rangle > f(\xi + u).$$

This implies that $\xi + u \in D$. Consider the real-valued function

$$\varphi(t) = f(\xi + tu) - f(\xi) - t\langle u, \xi^* \rangle$$

for $t \in [0, 1]$. Clearly φ is convex. Also $\varphi(0) = 0$, $\varphi(1) < 0$. Now φ has left and right derivatives for each t in $[0, 1]$. The right derivative φ'_r is nondecreasing and

$$\varphi(t) = \int_0^t \varphi'_r(s)\, ds.$$

Hence we can choose t_1 with $0 < t_1 < 1$ and $\varphi'_r(t) < 0$. Since $\tau = \xi + t_1 u$ belongs to D_f, we can find τ^* in $D_\tau f$ satisfying

$$f(\tau) + \langle v, \tau^* \rangle \le f(\tau + v)$$

for all v in X. Taking $v = (t - t_1)u$ we obtain

$$f(\tau) + (\tau - \tau_1)\langle u, \tau^* \rangle \le f(\tau + (t - t_1)u)$$

$$= f(\xi + tu)$$

for all real t, so that

$$\varphi(t) - \varphi(t_1) = f(\xi + tu) - f(\xi + t_1 u) - (t - t_1)\langle u, \xi^* \rangle$$

$$\geq (t - t_1)\langle u, \tau^* - \xi^* \rangle.$$

Since $\varphi'_r(t_1) < 0$, this implies that

$$\langle u, \tau^* - \xi^* \rangle < 0,$$

so that

$$\langle \tau - \xi, \tau^* - \xi^* \rangle = t_1 \langle u, \tau^* - \xi^* \rangle < 0$$

and (ξ, ξ^*) can be in the graph of no monotone map whose graph contains the point (τ, τ^*) in the graph of D_f.

Case (b) Suppose that $\xi \notin D$. Choose any point ζ in D and write $u = \xi - \zeta$. The line segment $\zeta + ut$, $0 \leq t \leq 1$ passes from the point ζ in D to the point ξ not in D. Since D is convex and open, we can choose t_1, $0 < t_1 \leq 1$, so that $\zeta + t_1 u$ lies on the boundary of D. Then $f(\zeta + t_1 u) = +\infty$ and $f(\zeta + tu)$ tends to $+\infty$ as t tends to t_1 from below. Since the function $\varphi(t) = f(\zeta + tu)$ is convex on $[0, t_1)$ we can choose t_2 with $0 < t_2 < t_1 \leq 1$ and

$$\varphi'_\ell(t_2) > \langle u, \xi^* \rangle.$$

Write $\tau = \zeta + t_2 u$ and choose τ^* in D_f. Then for all $\epsilon > 0$,

$$f(\tau) + \langle -\epsilon u, \tau^* \rangle \leq f(\tau - \epsilon u),$$

so that

$$\varphi(t_2 - \epsilon) - \varphi(t_2) = f(\tau - \epsilon u) - f(\tau) \geq -\epsilon \langle u, \tau^* \rangle$$

for $\epsilon > 0$. Thus

$$\lim_{\epsilon \to 0^+} \frac{\varphi(t_2 - \epsilon) - \varphi(t_2)}{-\epsilon} \leq \langle u, \tau^* \rangle$$

and

$$\langle u, \xi^* \rangle < \varphi'_\ell(t_2) \leq \langle u, \tau^* \rangle.$$

Now

$$\langle u, \xi^* - \tau^* \rangle < 0$$

and

$$\langle \xi - \tau, \xi^* - \tau^* \rangle = (1 - t_2)\langle u, \xi^* - \tau^* \rangle$$

$$< 0.$$

Again we conclude that (ξ, ξ^*) can be in the graph of no monotone map from X to X^* whose graph includes the point (τ, τ^*) in the graph of D_f.

Taking Cases (a) and (b) together we conclude that D_f is a maximal monotone map from X to X^*. $\quad\square$

Proof of Theorem 5.2. We first suppose that each weak* compact subset of X^* is weak* fragmented. Let D_f be the subdifferential map of a continuous convex function, from X to $\mathbb{R}_\infty$, that is finite for at least one point of X. Let D be the set of all points x where $f(x)$ is finite. By Lemma 5.3, D is an open convex set in X, and the subdifferential map D_f is a maximal monotone map from X to X^* with D as its domain.

By Theorem 5.1, the map D_f has a selector, h say, that is of the first Baire class, as a map from D to (X^*, norm), and is norm continuous at the points of a G_δ-set, say D_0, that is dense in D. This proves the first assertion of Theorem 5.2.

We now drop the assumption that the weak* compact subsets of X^* are weak* fragmented. We assume that the subdifferential map of each continuous convex function from X to $\mathbb{R}$ has a selector that is norm continuous at all points of a dense G_δ-set in X. Consider any such continuous convex function f from X to $\mathbb{R}$. We need to prove that f is Fréchet differentiable at all points of a G_δ-set dense in X. Using Lemma 5.3, the subdifferential map D_f is a maximal monotone map from X to X^* with X as its domain. By our special assumption we have a selector h for D_f on X that is continuous at all points of some G_δ-set, say D_0, dense in X. By Theorem 5.9, the set-valued function D_f is point-valued and norm upper semi-continuous at each point of D_0. Consider any point x_0 of D_0, and any point x of X. By the subdifferential property,

$$f(x_0) + \langle x - x_0, h(x_0)\rangle \le f(x),$$

$$f(x) + \langle x_0 - x, h(x)\rangle \le f(x_0),$$

so that $f(x)$ lies in the closed interval

$$[f(x_0) + \langle x - x_0, h(x_0)\rangle, f(x_0) + \langle x - x_0, h(x)\rangle]$$

of length

$$|\langle x - x_0, h(x) - h(x_0)\rangle|.$$

Thus

$$|f(x) - f(x_0) + \langle x - x_0, h(x_0)\rangle| \le |\langle x - x_0, h(x) - h(x_0)\rangle|$$

$$\le \|x - x_0\| \cdot \|h(x) - h(x_0)\|.$$

Since $h(x)$ converges in norm to $h(x_0)$ as x converges to x_0, the function f is

Fréchet differentiable with derivative $h(x_0)$ at x_0. Thus X is an Asplund space. □

5.3 ATTAINMENT MAPS FROM X^* TO X

In this section we give the proof of Theorem 5.3.

Proof of Theorem 5.3. Let K be a nonempty weakly compact set in the Banach space X. The attainment map from X^* to K is defined by

$$F_K(x^*) = \{x \in K : \langle x, x^* \rangle = \sup\{\langle k, x^* \rangle : k \in K\}\}$$

for all x^* in X^*. Since K is nonempty and weakly compact, we see that $F_K(x^*)$ is a nonempty weakly compact set, for each x^* in X^*. We regard X as a subset of X^{**} and prove that F_K is a monotone map from X^* to X^{**}. Let ξ^*, ζ^* be distinct points of X^* and suppose that ξ and ζ are points with

$$\xi \in F_K(\xi^*), \quad \zeta \in F_K(\zeta^*).$$

Then

$$\langle \xi, \xi^* \rangle = \sup\{\langle k, \xi^* \rangle : k \in K\} \geq \langle \zeta, \xi^* \rangle,$$

and similarly

$$\langle \zeta, \zeta^* \rangle \geq \langle \xi, \zeta^* \rangle,$$

so that

$$\langle \xi - \zeta, \xi^* - \zeta^* \rangle = \langle \xi, \xi^* \rangle - \langle \zeta, \xi^* \rangle + \langle \zeta, \zeta^* \rangle - \langle \xi, \zeta^* \rangle \geq 0.$$

Thus F_K is a monotone map from X^* to X^{**} that happens to take all its values in X.

We now verify that F_K is a weakly upper semi-continuous map from (X^*, norm) to (X, weak) taking only nonempty weakly compact values. Consider any ξ^* in X^* and any weakly open set G in X containing $F_K(\xi^*)$. Since $K \backslash G$ is weakly compact, $\langle x, \xi^* \rangle$ attains its supremum over $K \backslash G$ at some point h in $K \backslash G$. Since h is not in $F_K(\xi^*)$, this supremum, q say, must be strictly less than the value, p say, that $\langle x, \xi^* \rangle$ takes on $F_K(\xi^*)$. Thus

$$\langle x, \xi^* \rangle \leq q, \quad \text{for } x \text{ in } K \backslash G,$$

and

$$\langle x, \xi^* \rangle = p, \quad \text{for } x \text{ in } F_K(\xi^*).$$

Choose M so that

$$K \subset MB.$$

Then, provided

$$\|\zeta^* - \xi^*\| < \frac{p-q}{3M},$$

we have

$$|\langle x, \zeta^* \rangle - \langle x, \xi^* \rangle| \leq \frac{1}{3}(p-q)$$

for all x in K. Hence

$$\langle x, \zeta^* \rangle \leq \frac{2}{3}q + \frac{1}{3}p, \quad \text{for } x \text{ in } K \backslash G,$$

and

$$\langle x, \zeta^* \rangle \geq \frac{1}{3}q + \frac{2}{3}p, \quad \text{for } x \text{ in } F_K(\xi^*).$$

Thus $\langle x, \zeta^* \rangle$ cannot attain its supremum over K, which is at least $\frac{1}{3}q + \frac{2}{3}p$, at any point of $K \backslash G$. This implies that $F_K(\zeta^*) \subset G$. Thus F_K is weakly upper semi-continuous.

Now K, being weakly compact in X, is fragmented (see Namioka [60]). By Theorem 3.2, it follows that the weakly upper semi-continuous map F_K, taking only nonempty weakly compact values, has a selector, f say, of the first Baire class from (X^*, norm) to (X, norm), and f is norm to norm continuous on a dense G_δ-set U^* in X^*.

As we have seen, F_K is a monotone map from X^* to X^{**}. Let H be a maximal monotone map from X^* to X^{**} whose graph contains that of F_K. Then f is a selector for H that is norm to norm continuous at each point of U^*. By Theorem 5.1, the set of points, where H is point-valued and norm to norm upper semi-continuous, coincides with U^*. This implies that F_K is point-valued and norm to norm upper semi-continuous at each point of U^*. On the other hand, f has to be norm to norm continuous at each point where F_K is point-valued and norm to norm upper semi-continuous. Now consider any selector g for F_K. Then g is necessarily norm continuous at each point of U^*. Since g is also a selector for H, it follows, from the last part of Theorem 5.1, that each point where g is norm continuous belongs to U^*. $\quad \square$

5.4 ATTAINMENT MAPS FROM X TO X^*

In this section we give the proof of Theorem 5.4.

Proof of Theorem 5.4. Let K be a nonempty weak* compact set in the dual X^* of a Banach space X. The attainment map from X to K is defined by

$$F_K(x) = \{x^* \in K : \langle x, x^* \rangle = \sup\{\langle x, k^* \rangle : k^* \in K\}\}$$

for all x in X. The arguments used in the proof of Theorem 5.3 with the roles of

X and X^{**} taken over by X^* and the role of X^* taken over by X, show that F_K is a monotone map from X to X^*, with nonempty weak* compact values, and that F_K is weak* upper semi-continuous.

When K is convex and weak* fragmented by the norm on X^*, we can use Theorem 3.3 to deduce that F_K has a selector $f : X \rightarrow (K, \text{norm})$ that is of the first Baire class. $\quad\square$

5.5 METRIC PROJECTIONS OR NEAREST POINT MAPS

In this section we prove Theorems 5.5 and 5.6.

Proof of Theorem 5.5. We first prove that the nearest point map F from the Banach space X to its nonempty weakly compact set K is weakly upper semi-continuous. Recall that

$$\rho(x) = \inf\{\|x - k\| : k \in K\}$$

and

$$F(x) = \{k : \|x - k\| = \rho(x) \text{ and } k \in K\}.$$

We also use $\overline{B}(\xi; r)$ to denote the closed ball

$$\overline{B}(\xi; r) = \{x : \|x - \xi\| \leq r\}.$$

Since

$$F(x) = \bigcap_{n=1}^{\infty} K \cap \overline{B}(x; \rho(x) + (1/n))$$

is the intersection of a decreasing sequence of nonempty weakly compact sets, the set $F(x)$ is itself nonempty and weakly compact. Thus, for each x in X we can choose a point $k(x)$ in $F(x)$. Then, for each ξ in X we have

$$\rho(\xi) \leq \|\xi - k(x)\| \leq \|\xi - x\| + \|x - k(x)\|$$

$$= \|\xi - x\| + \rho(x).$$

Thus ρ satisfies the Lipschitz condition

$$|\rho(\xi) - \rho(x)| \leq \|\xi - x\|.$$

This implies that

$$F(\xi) \subset \overline{B}(\xi; \rho(\xi))$$

$$\subset \overline{B}(\xi; \rho(x) + \|\xi - x\|)$$

$$\subset \overline{B}(x; \rho(x) + 2\|\xi - x\|).$$

Now consider any weakly open set G containing $F(x)$. The sets

$$K \cap \overline{B}(x; \rho(x) + (1/n)) \backslash G,$$

$n \geq 1$, are a decreasing sequence of weakly compact sets with intersection

$$F(x) \backslash G = \emptyset.$$

Hence at least one of these sets, say

$$K \cap \overline{B}(x; \rho(x) + (1/n)) \backslash G,$$

is empty. Now, provided $\|\xi - x\| < (1/2n)$, we have

$$F(\xi) \subset K \cap \overline{B}(x; \rho(x) + 2\|\xi - x\|)$$

$$\subset K \cap \overline{B}(x; \rho(x) + (1/n)) \subset G.$$

Thus F is weakly upper semi-continuous at x, as required.

The weakly compact set K is automatically fragmented by the metric on X. Applying Theorem 3.2, it follows that F has a selector $f : X \rightarrow (K, \text{norm})$ that is of the first Baire class.

Now suppose that the norm on X is strictly convex. We show that F is a minimal weakly upper semi-continuous map with nonempty weakly compact values. Consider any weakly upper semi-continuous map H from X to K, taking only nonempty weakly compact values, with

$$H(x) \subset F(x)$$

for each x in X. Suppose that, for some x in X and some k in K we have

$$k \in F(x) \quad \text{and} \quad k \notin H(x).$$

Then

$$\|k - x\| = \rho(x).$$

Consider the point

$$\xi(\lambda) = x + \lambda(k - x)$$

for $0 < \lambda < 1$. Clearly

$$\rho(\xi(\lambda)) \leq \|k - \{x + \lambda(k - x)\}\|$$

$$= (1 - \lambda)\|k - x\| = (1 - \lambda)\rho(x).$$

If h is any point of K other than k we have

$$\|h - x\| \geq \rho(x).$$

Now

$$h - x = (h - \xi(\lambda)) + \lambda(k - x).$$

Since the norm is strictly convex we have

$$\|h - x\| \leq \|h - \xi(\lambda)\| + \lambda\|k - x\|$$

with strict inequality unless

$$h - \xi(\lambda) = (h - x) - \lambda(k - x)$$

is a positive scalar multiple of $k - x$, in which case $h - x$ is a positive scalar multiple of $k - x$. So we have

$$\|h - \xi(\lambda)\| \geq \|h - x\| - \lambda\|k - x\|$$

$$\geq \rho(x) - \lambda\rho(x)$$

$$= (1 - \lambda)\rho(x) \geq \rho(\xi(\lambda)),$$

with strict inequality unless $h - x$ is a positive scalar multiple of $k - x$. In both cases we conclude that

$$\|h - \xi(\lambda)\| > \rho(\xi(\lambda)).$$

Thus h does not belong to $F(\xi(\lambda))$, and $F(\xi(\lambda))$ reduces to the single point k. Since

$$\emptyset \neq H(\xi(\lambda)) \subset F(\xi(\lambda)) = \{k\},$$

the set $H(\xi(\lambda))$ also reduces to the point k. If we now let λ tend to zero and use the weak lower semi-continuity of H at x, we conclude that $k \in H(x)$. This contradiction shows that F is minimal.

We now apply Theorem 3.5, remembering that a Banach space is complete, and obtain a dense G_δ-set U in X, with F point-valued and norm upper semi-continuous at each point of U. $\square$

Proof of Theorem 5.6. The proof follows the proof of Theorem 5.5 with a few minor changes. $\square$

5.6 SOME SELECTIONS INTO FAMILIES OF CONVEX SETS

In this section we prove Theorems 5.7 and 5.8 obtaining them as consequences of a more complicated theorem. We need to introduce a special type of joint continuity.

If X is a topological space and (Y, ρ) is a metric space we say that *a map $\ell : X \times Y \to \mathbb{R}$ is jointly (a) continuous with respect to X and (b) uniformly continuous on bounded sets with respect to Y, if, given any $\epsilon > 0$, any x^* in X and any bounded set S in Y, there is a neighborhood N of x^* in X and a $\delta > 0$, with*

$$|\ell(x,y) - \ell(x^*, y^*)| < \epsilon,$$

whenever $x \in N$ and y, y^* in S satisfy $\rho(y, y^*) < \delta$.

We note that this condition is satisfied by the function

$$\ell(x, y) = \|y - \eta(x)\|$$

under the conditions of Theorem 5.7, and by the function

$$\ell(x, y) = -\langle y, \eta^*(x)\rangle$$

under the conditions of Theorem 5.8. Further, in each of these cases, for each x in X and each real t, the set of y satisfying

$$\ell(x, y) \le t$$

is weakly closed in Y. Thus, in each case, $\ell(x, \cdot)$ is a weakly lower semi-continuous real-valued function.

We prove the following theorem, and deduce Theorems 5.7 and 5.8 directly from it.

Theorem 5.10 *Let X be a metric space and let Y be a Banach space that is σ-fragmented. Let H be a set-valued map from X to Y with nonempty convex weakly compact values, and suppose that H is a continuous map from X to the space of nonempty bounded norm closed sets in Y taken with the Hausdorff metric.*

Let $\ell : X \times Y \to \mathbb{R}$ be a map that is jointly (a) continuous with respect to X and (b) uniformly continuous on bounded sets with respect to Y. Suppose further that, for each x in X, $\ell(x, \cdot)$ is lower semi-continuous as a real-valued function on $(Y, weak)$.

Write

$$m(x) = \inf\{\ell(x, y) : y \in H(x)\}$$

for each x in X. Then it is possible to choose a sequence $h_1, h_2, \ldots$ of continuous selectors for H converging pointwise to a selector h for H satisfying

$$\ell(x, h(x)) = m(x)$$

for each x in X.

We prove three lemmas.

Lemma 5.4 *Let Y be a Banach space. Let K be a weakly compact set contained in a weakly open set G in Y. Then there is an $\epsilon > 0$ and a weakly open set J with*

$$K \subset J \subset G$$

and with $\|j - e\| \geq \epsilon$ *whenever* $j \in J$ *and* $e \notin G$.

Proof. Since K is a weakly compact subset of the weakly open set G, there is a weak neighborhood U of the origin of Y such that $K + U \subset G$ (see, e.g., Kelley and Namioka [44, Theorem 5.2 (vi), p. 35]). Choose $\epsilon > 0$ so that

$$B(0; \epsilon) = \{y : \|y\| < \epsilon\}$$

is contained in $\frac{1}{2} U$. Then $J = K + \frac{1}{2} U$ is weakly open and

$$J + B(0; \epsilon) \subset J + \frac{1}{2} U \subset K + U \subset G.$$

The result follows. $\square$

Lemma 5.5   *Let X be a metric space and let Y be a Banach sphere. Let H be a set-valued map from X to Y with nonempty weakly compact values and suppose that H is a continuous function from X to the space of nonempty bounded norm closed sets in Y taken with the Hausdorff metric.*

Let $\ell : X \times Y \to \mathbb{R}$ be a map that is jointly (a) continuous with respect to X and (b) uniformly continuous on bounded sets with respect to Y. Suppose further that, for each x in X, $\ell(x, \cdot)$ is lower semi-continuous as a real-valued function on (Y, weak).

Write

$$m(x) = \inf\{\ell(x, y) : y \in H(x)\}$$

and

$$F(x) = H(x) \cap \{y : \ell(x, y) = m(x)\}$$

for each x in X.

Then m is a continuous real-valued function on X and F is a weakly upper semi-continuous set-valued function from X to Y taking only nonempty weakly compact values.

Proof. Since $H(x)$ is nonempty and weakly compact and $\ell(x, \cdot)$ is a weakly lower semi-continuous real-valued function, the infimum $m(x)$ of $\ell(x, \cdot)$ over $H(x)$ is attained on $H(x)$; let $\eta(x)$ be some point of $H(x)$ where this infimum is attained. Note also that $F(x) = H(x) \cap \{y : \ell(x, y) = m(x)\}$ is weakly compact.

We first prove that m is continuous on X. Let x^* in X and $\epsilon > 0$ be given. Then

$$H(x^*) + B(0; 1)$$

is bounded in Y. By the continuity property of ℓ, we can choose a neighborhood N_1 of x^* and a δ with $0 < \delta < 1$, so that

$$|\ell(x, y) - \ell(x^*, y^*)| < \epsilon,$$

whenever x is in N_1 and y, y^* are in $H(x^*) + B(0; 1)$ and satisfy $\|y - y^*\| < \delta$. By the continuity of H in the Hausdorff metric, d say, we can choose a neighborhood N_2 of x^* such that

$$d(H(x), H(x^*)) < \delta$$

for all x in N_2. Now, provided $x \in N_1 \cap N_2$, for each y in $H(x)$ there is a y^* in $H(x^*)$ with $\|y - y^*\| < \delta$, so that

$$\ell(x, y) \geq \ell(x^*, y^*) - \epsilon$$

$$\geq m(x^*) - \epsilon.$$

Hence $m(x) \geq m(x^*) - \epsilon$, when $x \in N_1 \cap N_2$. Further, when $x \in N_1 \cap N_2$, there will be an η in $H(x)$ with $\|\eta - \eta(x^*)\| < \delta$, so that

$$m(x) \leq \ell(x, \eta)$$

$$\leq \ell(x^*, \eta(x^*)) + \epsilon$$

$$= m(x^*) + \epsilon.$$

Thus m is continuous on X.

It remains to prove that F is weakly upper semi-continuous. Let x^* be any point of X and let G be any weakly open set in Y that contains $F(x^*)$. We need to show that $F(x) \subset G$ for all x in a suitable neighborhood of x^*. We suppose that this is not the case and we seek a contradiction. Then, for each neighborhood N of x^*, there is a point $x(N)$ in N with

$$F(x(N)) \backslash G \neq \emptyset.$$

By Lemma 5.4, we can choose $\epsilon > 0$ and a weakly open set J with

$$F(x^*) \subset J \subset G$$

and with $\|j - e\|$, for all points j in J and e not in G.

By the continuity property of ℓ, for each $n \geq 1$, we can choose δ_n with $0 < \delta_n < \epsilon$ and a neighborhood $N_n^{(1)}$ of x^* with

$$|\ell(x, y) - \ell(x^*, y^*)| < 1/n,$$

whenever x is in $N_n^{(1)}$ and y, y^* are in $H(x^*) + B(0; 1)$ with $\|y - y^*\| < \delta_n$.

By the continuity of H we can choose a neighborhood $N_n^{(2)}$ of x^* with

$$d(H(x), H(x^*)) < \delta_n$$

for all x in $N_n^{(2)}$.

By the continuity of m, we can choose a neighborhood $N_n^{(3)}$ of x^* such that

$$|m(x) - m(x^*)| < 1/n$$

for all x in $N_n^{(3)}$.

Using our supposition concerning the neighborhoods N of x^*, for each $n \geq 1$, we can choose x_n in $N_n = N_n^{(1)} \cap N_n^{(2)} \cap N_n^{(3)}$ and y_n in Y with

$$y_n \in F(x_n), \quad y_n \notin G.$$

Since $x_n \in N_n^{(2)}$, we can choose z_n in $H(x^*)$ with

$$\|z_n - y_n\| \leq d(H(x_n), H(x^*)) < \delta_n < \epsilon.$$

Since $y_n \notin G$, it follows from our choice of J, that

$$z_n \notin J.$$

Since $x_n \in N_n^{(1)}$ and z_n and y_n belong to $H(x^*) + B(0; 1)$ and satisfy $\|z_n - y_n\| < \delta_n$, we have

$$\ell(x^*, z_n) \leq \ell(x_n, y_n) + 1/n.$$

Since $y_n \in F(x_n)$ and $x_n \in N_n^{(3)}$, this yields

$$\ell(x^*, z_n) \leq m(x_n) + 1/n$$

$$\leq m(x^*) + 2/n.$$

Thus

$$\ell(x^*, y_n) \to m(x^*)$$

as $n \to \infty$.

For $n \geq 1$, write

$$Z_n = \text{wk cl}\{z_n, z_{n+1}, z_{n+2}, \ldots\}.$$

Then, for each $n \geq 1$ and $\zeta > 0$, the set

$$(H(x^*) \backslash J) \cap \{y : \ell(x^*, y) \leq m(x^*) + \zeta\} \cap Z_n$$

is a weakly compact set in Y containing z_m for all sufficiently large m. Since these sets decrease as n increases and ζ decreases, the set

$$\bigcap_{\eta > 0} \bigcap_{n \geq 1} (H(x^*) \backslash J) \cap \{y : \ell(x^*, y) \leq m(x^*) + \zeta\} \cap Z_n$$

is nonempty. Take z to be a point in this set. Then

$$z \in (H(x^*) \backslash J) \cap \{y : \ell(x^*, y) \leq m(x^*)\}$$

$$\subset F(x^*) \backslash J = \emptyset.$$

This contradiction completes the proof. $\quad\square$

Lemma 5.6 *Let X be a metric space and let Y be a Banach space. Let H be a norm lower semi-continuous map from X to Y taking only nonempty convex norm closed values. Let h be a selector for H of the first Baire class. Then there is a sequence $h_1, h_2, h_3, \ldots$ of continuous selectors for H that converge pointwise to h.*

Proof. Since h is of the first Baire class, we can choose a sequence $f_1, f_2, f_3, \ldots$ of continuous functions from X to Y converging pointwise to h. We first show that there is a second such sequence $g_1, g_2, g_3, \ldots$ converging pointwise to h and satisfying

$$d(g_n(x), H(x)) < 1/n$$

for all x in X, where we use $d(y, H)$ to denote

$$\inf\{\|y - h\| : h \in H\}.$$

For each n, m with $m \geq n \geq 1$, write

$$U_{nm} = \{x \in X : d(f_m(x), H(x)) < 1/n\}.$$

If $x \in U_{nm}$, then there is a point η in $H(x)$ with

$$\|f_m(x) - \eta\| < 1/n.$$

Write

$$\epsilon = (1/n) - \|f_m(x) - \eta\|.$$

Since f_m is continuous and H is lower semi-continuous, we can choose $\delta > 0$ so that

$$\|f_m(\xi) - f_m(x)\| < \frac{1}{2}\epsilon$$

and there is a point $\zeta = \zeta(\xi)$ in $H(\xi)$ with

$$\|\zeta - \eta\| < \frac{1}{2}\epsilon$$

for all ξ with $\|\xi - x\| < \delta$. Now, for each ξ with $\|\xi - x\| < \delta$,

$$\begin{aligned}
d(f_m(\xi), H(\xi)) &\leq \|f_m(\xi) - \zeta\| \\
&\leq \|f_m(x) - \eta\| + \|\eta - \zeta\| + \|f_m(\xi) - f_m(x)\| \\
&< 1/n,
\end{aligned}$$

and so $\xi \in U_{nm}$. Thus U_{nm} is open.

Since f_m converges pointwise to the selector h for H, the family

$$\{U_{nm} : m \geq n\}$$

covers X, for each fixed $n \geq 1$. By Lemma 1.1, for each $n \geq 1$, there will be a locally finite partition of unity $\{p_{n\gamma} : \gamma \in \Gamma(n)\}$ for X refining the family

$\{U_{nm} : m \geq n\}$. For each γ in $\Gamma(n)$, choose $m(\gamma) \geq n$ with the support of $p_{n\gamma}$ contained in $U_{nm}(\gamma)$. Write

$$g_n(x) = \sum \{p_{n\gamma}(x)f_{m(\gamma)}(x) : \gamma \in \Gamma(n)\}$$

for $n \geq 1$ and $x \in X$. Then each g_n is continuous. Since the support of $p_{n\gamma}(x)$ is contained $U_{nm}(\gamma)$, we have

$$d(f_{m(\gamma)}(x), H(x)) < 1/n,$$

when $p_{n\gamma}(x) \neq 0$. Using the convexity of $H(x)$ this yields

$$d(g_n(x), H(x)) < 1/n$$

for all x in X and $n \geq 1$. To see that g_n converges pointwise to h, let x in X and $\epsilon > 0$ be given. Choose N so large that $\|f_n(x) - h(x)\| < \epsilon$ for $n \geq N$. Then, for $n \geq N$,

$$\|g_n(x) - h(x)\| = \left\| \sum \{p_{n\gamma}(x)\left(f_{m(\gamma)}(x) - h(x)\right) : \gamma \in \Gamma(n)\right\|$$

$$\leq \sum \{p_{n\gamma}(x)\|f_{m(\gamma)}(x) - h(x)\| : \gamma \in \Gamma(n)\}$$

$$< \epsilon$$

as required.

Now, for each $n \geq 1$, and each x in X, write

$$L_n(x) = B(g_n(x), 1/n) \cap H(x)$$

and

$$M_n(x) = \text{norm cl } L_n(x).$$

The condition $d(g_n(x), H(x)) < 1/n$ ensures that $L_n(x)$ is nonempty. To prove that L_n is a norm lower semi-continuous map, for each $n \geq 1$, we use an argument similar to that used to prove that the sets U_{mn} are open. Consider any $n \geq 1$, any x in X and any y in $L_n(x)$. Then there is a point η in $H(x)$ with

$$\|g_n(x) - \eta\| < 1/n.$$

Write

$$\epsilon = (1/n) - \|g_n(x) - \eta\|.$$

Using the continuity of $g_n(x)$ and the norm lower semi-continuity of H, we can choose $\delta > 0$, so that

$$\|g_n(\xi) - g_n(x)\| < \frac{1}{2}\epsilon$$

and there is a point $\zeta = \zeta(\xi)$ in $H(\xi)$ with

$$\|\zeta - \eta\| < \frac{1}{2}\epsilon$$

for all ξ with $\|\xi - x\| < \delta$. Then

$$\|\zeta - g_n(\xi)\| \le \|\eta - g_n(x)\| + \|\zeta - \eta\| + \|g_n(\xi) - g_n(x)\|$$

$$< 1/n$$

for all ξ with $\|\xi - x\| < \delta$. Thus, for all ξ with $\|\xi - x\| < \delta$ there is point ζ in $L_n(\xi)$ with $\|\zeta - \eta\| < \frac{1}{2}\epsilon$. Since we may replace ϵ by any smaller positive number, it follows that L_n is norm lower semi-continuous for each $n \ge 1$. Clearly the norm closure M_n of L_n is also norm lower semi-continuous.

Since M_n maps to the nonempty closed convex sets of Y, Michael's selection theorem, Theorem 1.1, ensures the existence of a continuous selector h_n for M_n. Now h_n is a continuous selector for H satisfying

$$\|h_n(x) - g_n(x)\| \le 1/n$$

for all x. Hence the sequence h_1, h_2, h_3, converges pointwise to h and satisfies our requirements. □

Proof of Theorem 5.10. We write

$$F(x) = H(x) \cap \{y : \ell(x, y) = m(x)\}$$

for each x in X. By Lemma 5.5, the map F is a weakly upper semi-continuous set-valued map taking only nonempty weakly compact values. By Theorem 4.2 part (d), the set-valued map F has a selector h of the first Baire class.

Since H is continuous in the Hausdorff metric, it is norm lower semi-continuous as a set-valued map to Y. (Note that, in general, it does not follow that this map is norm upper semi-continuous.) Now, by Lemma 5.6, we can choose the required continuous selectors $h_1, h_2, h_3, \ldots$ for H converging to the first Baire class selector h for H satisfying

$$\ell(x, h(x)) = m(x)$$

for each x in X. □

Proof of Theorem 5.7. The result follows directly from Theorem 5.10 and the remarks before the statement of that theorem. □

Proof of Theorem 5.8. The result follows directly from Theorem 5.10 and the remarks before the statement of that theorem. □

5.7 EXAMPLE

We give an example illustrating the difficulties that arise when one studies lower semi-continuous convex functions or maximal monotone maps whose domain contains no interior point.

Example 5.1 *There is a lower semi-continuous convex function $f : \ell^2 \to \mathbb{R}_\infty$ with the following properties. The set L_1 of points of ℓ^2 where f is finite is a dense linear subspace of ℓ^2 that is not closed. The subdifferential D_f is well defined; the set L_2 of points where D_f is nonempty is a relatively dense linear subspace of L_2 strictly contained in L_1. D_f is a maximal monotone map on ℓ^2. At each point x of L_2 the set $D_f(x)$ reduces to a single point, say $f'(x)$. The function f' is locally bounded at no point of L_2; it is, however, of the first Baire class on L_2.*

Construction. Let

$$x = x_1, x_2, \ldots,$$

where

$$\|x\| = \sqrt{\sum_{i=1}^{\infty} x_i^2} < +\infty$$

denotes a typical point of ℓ^2 and its norm. Define $f : \ell^2 \to \mathbb{R}_\infty$ by

$$f(x) = \sum_{i=1}^{\infty} i x_i^2.$$

Clearly the set L_1 of all points x of ℓ^2 for which $f(x)$ is finite is a dense linear subspace of ℓ^2 that does not coincide with ℓ^2. Further f is convex on ℓ^2. For each real t the set of points x of ℓ^2 with

$$f(x) = \sum_{i=1}^{\infty} i x_i^2 \leq t \tag{5.1}$$

is closed in ℓ^2. To see this, suppose that a sequence of points $x^{(1)}, x^{(2)}, x^{(3)}, \ldots$ of ℓ^2 satisfying (5.1) converges to a point $x^{(0)}$ of ℓ^2. Then

$$\sum_{i=1}^{\infty} i \left(x_i^{(0)} \right)^2 = \sup_N \sum_{i=1}^{N} i \left(x_i^{(0)} \right)^2$$

$$= \sup_N \lim_{n \to \infty} \sum_{i=1}^{N} i \left(x_i^{(n)} \right)^2$$

$$\leq t$$

and $x^{(0)}$ belongs to the set. Thus f is lower semi-continuous. Now let L_2 be the set of all points x of ℓ^2 with

$$\sum_{i=1}^{\infty} i^2 x_i^2 < +\infty.$$

Clearly L_2 is a relatively dense linear subspace of L_1 strictly contained in L_1. We want to determine the subdifferential D_f of f. If $x \notin L_1$, then $f(x) = +\infty$ and $D_f(x) = \emptyset$. Now consider any x in L_1 and any d in ℓ_2. Suppose that we can find an $i \geq 1$ with

$$d_i \neq 2ix_i.$$

Then we can find a point $u^* = u_1^*, u_2^*, u_3^*, \ldots$ in ℓ^2 with

$$u_j^* = 0, \quad \text{when } j \neq i,$$

$$0 < |iu_i^*| < |2ix_i - d_i|, \text{ and}$$

$$\text{sign } u_i^* = \text{sign}(d_i - 2ix_i).$$

Then

$$\text{sign}(d_i - 2ix_i - iu_i^*) = \text{sign}(d_i - 2ix_i) = \text{sign } u_i^*.$$

Thus

$$(d_i - 2ix_i - iu_i^*)u_i^* > 0$$

and

$$u_i d_i > 2ix_i u_i^* + iu_i^{*2},$$

so that

$$f(x) + \langle u^*, d \rangle = \left(\sum_{j=1}^{\infty} jx_j^2\right) + u_i d_i$$

$$> f(x + u^*).$$

This ensures that $d \notin D_f(x)$. Hence the only point that can possibly be in $D_f(x)$ is

$$2(x_1, 2x_2, 3x_3, \ldots) = f'(x),$$

say. However, this point is not in ℓ^2 when $x \in L_1 \backslash L_2$. So

$$D_f(x) = \emptyset, \quad \text{when } x \notin L_2,$$

$$D_f(x) \subset \{f'(x)\}, \quad \text{when } x \in L_2.$$

It is easy to verify that

$$f(x) + \langle u, f'(x) \rangle \leq f(x + u)$$

for all u in ℓ^2 when $x \in L_2$. Hence

$$D_f(x) = \{f'(x)\}, \quad \text{when } x \in L_2.$$

We now verify that D_f is a maximal monotone map on ℓ_2. If $x_1, x_2 \in \ell^2$ and

$$x_1^* \in D_f(x_1), \quad x_2^* \in D_f(x_2),$$

then x_1 and x_2 belong to L_2 and

$$x_1^* = f'(x_1), \quad x_2^* = f'(x_2),$$

so that

$$\langle x_2 - x_1, x_2^* - x_1^* \rangle = \langle x_2 - x_1, f'(x_2) - f'(x_1) \rangle$$

$$= \sum_{i=1}^{\infty} \left(x_i^{(2)} - x_i^{(1)} \right) \left(2ix_i^{(2)} - 2ix_i^{(1)} \right)$$

$$= 2 \sum_{i=1}^{\infty} \left(x_i^{(2)} - x_i^{(1)} \right)^2 \geq 0.$$

Thus D_f is a monotone map.

Suppose that D_f is not a maximal monotone map, so that there is a monotone map H with

$$H(x) \supset D_f(x), \quad \text{for all } x,$$

and

$$H(\xi) \backslash D_f(\xi) \neq \emptyset$$

for some ξ in ℓ_2. Then we can choose ζ in

$$H(\xi) \backslash D_f(\xi).$$

If $\xi \in L_2$, then

$$\sum_{i=1}^{\infty} i^2 \xi_i^2 < +\infty$$

and $f'(\xi)$ is welldefined. Since $\zeta \notin D_f(\xi)$ we have

$$\zeta \neq f'(\xi)$$

and we can choose $i \geq 1$ with

$$\zeta_i \neq 2i\xi_i.$$

Similarly, if $\xi \notin L_2$, we have

$$\sum_{i=1}^{\infty} i^2 \xi_i^2 = +\infty,$$

$$\sum_{i=1}^{\infty} \zeta_i^2 < +\infty,$$

so that we can again choose $i \geq 1$, with

$$\zeta_i \neq 2i\xi_i.$$

In each case write

$$\xi' = \xi_1, \xi_2, ..., \xi_i + \delta_i, \xi_{i+1}, ...$$

with δ_i to be chosen later. Then

$$\zeta \in H(\xi)$$

and ζ', defined by $\zeta' = f'(\xi')$, belongs to

$$D_f(\xi') \subset H(\xi').$$

But

$$\langle \xi' - \xi, \zeta' - \zeta \rangle = (\xi_i + \delta_i - \xi_i)(2i\xi_i + 2i\delta_i - \zeta_i)$$

$$= \delta_i(2i\xi_i - \zeta_i + 2i\delta_i)$$

$$< 0,$$

provided δ_i is sufficiently small with the same sign as $\zeta_i - 2i\xi_i$. Thus H cannot be a monotone map and D_f is a maximal monotone map.

The function f' is clearly a selector for the maximal monotone map D_f on its domain L_2. However, if $\xi \in L_2$ and we take

$$\xi^{(i)} = \xi + (0, 0, 0, ..., 1/\sqrt{i}, ...),$$

the nonzero element $1/\sqrt{i}$ occurring in the ith place, we see that $\xi^{(i)}$ converges to ξ through L_2 as $i \to \infty$, but

$$f'(\xi^{(i)}) = f'(\xi) + (0, 0, 0, ..., 2\sqrt{i}, ...),$$

so that

$$\|f'(\xi^{(i)})\| \to \infty \quad \text{as } i \to \infty.$$

On the other hand, f' is the pointwise limit on L_2 of the sequence of continuous functions $h^{(i)}$, $i = 1, 2, 3, ...$, defined by

$$h^{(i)}(x) = (2x_1, 4x_2, 6x_3, \ldots, 2ix_i, 0, 0, \ldots)$$

for $x \in L_2$.

5.8 REMARKS

The results in Sections 5.1–5.5 are refinements of some results in a field that has been much investigated. The results on monotone maps, maximal monotone maps and subdifferentials were largely developed from simpler beginnings by R. T. Rockafellar. The results on attainment maps and the nearest point map owe much to P. S. Kenderov. For an account of much of this work see Phelps [66]. However, all these writers ignore the existence of selectors of the first Baire class; such selectors were first found by Jayne and Rogers [31].

Although we have obtained Theorem 5.3, on the attainment map $F_K : X^* \to K$ from the dual X^* of a Banach space X to a weakly compact set K in X, another approach is perhaps simpler in the special case when K is convex. In this case, it is easy to prove directly that F_K is a weakly upper semi-continuous map from X^* to K, taking only nonempty weakly compact *convex* values. One can then prove that F_K is minimal amongst all such maps taking only nonempty weakly compact convex values. Such a minimal map is necessarily point-valued and norm upper semi-continuous on a norm dense G_δ-set U in X^*. The map F_K being weakly upper semi-continuous with nonempty weakly compact values has a selector f, of the first norm Baire class, which is necessarily norm continuous at the points of U.

When we have proved Srivatsa's theorem, see Theorem 6.2 below, we shall be able to drop the conditions in Theorems 5.7 and 5.8 that the Banach space Y be σ-fragmented. These theorems appear here for the first time. We are grateful to E. Michael and I. Namioka for some comments on a previous draft.

Chapter 6

Selectors for upper semi-continuous set-valued maps with nonempty values that are otherwise arbitrary

In this chapter we prove three main theorems, the first two due to V. V. Srivatsa, and the third by use of his method (see [74]). Srivatsa obtained his results and wrote them out in detail during a visit to University College London for the session 1984–85; publication was much delayed.

Another account is given by Jayne, Orihuela, Pallarés and Vera [41].

Theorem 6.1 (Srivatsa) *Let X and Y be metric spaces and let F be an upper semi-continuous set-valued map from X to Y taking only nonempty values. Then F has a selector f that is σ-discrete and of the first Borel class.*

Theorem 6.2 (Srivatsa) *Let X be a metric space and let Z be a convex subset of a Banach space Y. Let F be an upper semi-continuous set-valued map from X to Z, with the weak topology of Y, taking only nonempty values. Then F has a selector f, which, when regarded as a map from X to Z with the norm topology of Y, is of the first Baire class.*

Theorem 6.3 (After Srivatsa) *Let X be a metric space and let K be a compact Hausdorff space. Let Z be a convex subset of the space $C(K)$ of continuous real-valued functions on K. Let F be an upper semi-continuous set-valued map from X to Z, with the topology of pointwise convergence of $C(K)$, taking only nonempty values. Then F has a selector f, which, when regarded as a map from X to Z with the topology of uniform convergence on $C(X)$, is of the first Baire class.*

We also prove the following result of Srivatsa, extending a result of Kuratowski and Ryll-Nardzewski (see Chapter 1, Remark 10).

Theorem 6.4 *Let X and Y be metric spaces. Let F be a lower semi-continuous set-valued map from X to Y taking only nonempty values that are complete in the metric of Y. Then F has a selector f that is σ-discrete and of the first Borel class.*

In Section 6.1 we prove some "diagonal lemmas". Although we shall not make explicit use of the concept, these lemmas are related to the "boundary" of an upper semi-continuous map, introduced by Choquet [6], and developed by S. Dolecki, S. Rolewicz, A. Lechicki, and J. E. Jayne, C. A. Rogers, R. W. Hansell and I. Labuda; see the paper [21] by the four last named authors and the references given there.

In Section 6.2 we give the proofs of the main theorems. In Section 6.3 we prove Theorem 6.4. In view of Theorems 6.2 and 6.3 and recalling Theorem 3.3, it is perhaps suprising that we cannot prove the following apparently relatively weak statement.

(A)   Let X be a complete metric space and let Y^* be a dual Banach space that is weak* σ-fragmented using weak* closed sets. Let F be an upper semi-continuous set-valued map from X to Y^* with its weak* topology, taking only nonempty norm complete values. Then F has a selector f that is of the first Baire class as a map from X to Y^* with its norm topology.

Of course, if the words "weak* compact" were substituted for the words "norm complete", the modified statment would follow from Theorem 3.3. Never-the-less, in Section 6.4. we give an example, suggested to us in detail by M. Valdivia, showing that the statement (A) is false.

In Section 6.5 we make some remarks.

6.1 DIAGONAL LEMMAS

If X is a metric space and Y is a topological space and F is a set-valued map from X to Y, we say that F has the *diagonal property* if: whenever x_0 is the limit of a sequence $x_1, x_2, \ldots$ of points of X, all distinct from x_0, and

$$y_i \in F(x_i) \backslash F(x_0), \quad i \geq 1,$$

then there is a point y_0 in $F(x_0)$ that is the limit point of a subsequence of the sequence $y_1, y_2, y_3, \ldots$. A simple example of a set-valued map with this property is given in Remark 5 at the end of this chapter. If X is a metric space and Z is a convex set in a normed linear space Y, and F is a set-valued function from X to Z, we say that F has the *convex diagonal property* if: whenever x_0 is the limit of a sequence $x_1, x_2, \ldots$ of points, all distinct from x_0, and

$$y_i \in F(x_i) \backslash F(x_0), \quad i \geq 1,$$

then there is a point y_0 in $F(x_0)$ that is the norm limit of a sequence of finite rational convex combinations of the points $y_i, i \geq 1$.

Lemma 6.1   *Let X and Y be metric spaces. Let F be a set-valued map from X to Y with $F^{-1}(C)$ closed in X whenever C is a closed countable set in Y. Then F has the diagonal property.*

Proof. Let x_0 be the limit of a sequence $x_1, x_2, \ldots$ of points of X, all distinct from x_0. Suppose that

$$y_i \in F(x_i) \backslash F(x_0), \quad i \geq 1.$$

We need to show that there is a point y_0 in $F(x_0)$ that is the limit of a subsequence of the sequence $y_1, y_2, y_3, \ldots$. If the set

$$C = \{y_i : i \geq 1\}$$

were closed in Y, the set

$$F^{-1}(C)$$

would be closed in X, and containing the points $x_1, x_2, \ldots$, would also contain x_0, ensuring that at least one of the points y_i, $i \geq 1$, of C would lie in $F(x_0)$, contradicting the choice of $y_1, y_2, y_3, \ldots$. Hence the set C is not closed.

Now there will be a point, z_0 say, not in C, and a sequence $z_1, z_2, z_3, \ldots$ of distinct points chosen from $y_1, y_2, y_3, \ldots$ converging to z_0. The set

$$D = \{z_i : i \geq 0\}$$

is a closed countable set in Y. Thus $F^{-1}(D)$ is closed in X and so contains x_0. Since $F(x_0)$ contains none of the points z_i, $i \geq 1$, it must contain z_0. Now the point $y_0 = z_0$ satisfies our requirements. $\square$

Lemma 6.2 *Let X be a metric space and let Z be a convex set in a Banach space Y. Let F be a set-valued map from X to Z with $F^{-1}(C)$ closed in X whenever C is closed separable and convex in Z using the norm topology of Y. Then F has the convex diagonal property.*

Proof. Suppose that x_0 is the limit of a sequence $x_1, x_2, \ldots$ of points of X, all distinct from x_0, and that

$$y_i \in F(x_i) \backslash F(x_0), \quad i \geq 1.$$

We need to show that there is a point y_0 in $F(x_0)$ that is the limit in norm of a sequence of finite rational convex combinations of the points y_i, $i \geq 1$.

Let C be the norm closure in Z of the finite rational convex combinations of points chosen from $\{y_i : i \geq 1\}$. Then C is closed, separable and convex in Z. Hence $F^{-1}(C)$ is closed in X. Since

$$x_i \in F^{-1}(C), \quad i \geq 1,$$

we must have

$$x_0 \in F^{-1}(C)$$

and there is a point y_0 of C in $F(x_0)$. Consequently, y_0 in $F(x_0)$ is the norm limit of a sequence of finite rational convex combinations of points chosen from $\{y_i : i \geq 1\}$, as required $\square$.

Before our next lemmas, we give some definitions and quote some results. A convenient reference is Floret's book [12]; we give some page references to that book. A set A in a Hausdorff space X is said to be *countably compact* if every sequence in A has a cluster point in A, and A is said to be *relatively countably compact* if every sequence in A has a cluster point in X (see p. 7). A Hausdorff space X is said to be *angelic*, if each relatively countably compact set A in X is relatively compact in X and each point of A's closure is the limit of a sequence of points in A (see p. 30). It is known that when K is a compact Hausdorff space, the space $C_p(K)$ of continuous real-valued functions on K, with the topology of pointwise convergence, is angelic (see p. 36, referring back to p. 11 for the definition of ω_X). By a result of Grothendieck (see p. 45), a set A in the Banach space $C(K)$ is weakly compact, if, and only if, it is pointwise compact and bounded. Further (see p. 47), when A in $C(K)$ is weakly compact, the weak and the pointwise topologies coincide on A.

Lemma 6.3 *Let X be a metric space and let Y be a Hausdorff space that is angelic. Let F be a set-valued map from X to Y with $F^{-1}(C)$ closed in X whenever C is closed and separable in Y. Then F has the diagonal property.*

Proof. Let x_0 be the limit point of a sequence $x_1, x_2, \ldots$ of points of X, all distinct from x_0. Suppose that $y_1, y_2, y_3, \ldots$ is a sequence of points of Y with

$$y_i \in F(x_i)\backslash F(x_0), \quad i \geq 1.$$

We need to show that there is a point y_0 in $F(x_0)$ that is the limit of a subsequence of the sequence $y_1, y_2, y_3, \ldots$.

Write

$$C = \mathrm{cl}\{y_i : i \geq 1\}.$$

Then $F^{-1}(C)$ is closed in X, contains the points $x_1, x_2, \ldots$, and so also contains x_0. Thus $F(x_0)$ contains a point y_0 that is not in the set

$$S = \{y_i : i \geq 1\},$$

but is in the closure C of this set. In particular, S has a cluster point in Y.

Applying this argument to each subsequence of the sequence (y_i), we find that S is relatively countably compact. Since Y is angelic, S is relatively compact in Y and, by the remark above, each point of the closure C of S is the limit of a sequence of points of S. Thus the point y_0 chosen in the last paragraph satisfies our requirements. $\square$

Lemma 6.4 *Let K be a compact Hausdorff space and let $C(K)$ be the Banach space of real-valued continuous functions on K. Let X be a metric space and* *let Z be a bounded convex set in $C(K)$. Let F be a pointwise upper semi-continuous set-valued map from X to $C(K)$, taking values in Z. Then F has the convex diagonal property.*

Proof. Since $C_p(K)$ is angelic, the hypotheses of Lemma 6.3 are satisfied with $Y = C_p(K)$. Thus F, regarded as a map to $C_p(K)$, has the diagonal property.

Now suppose that x_0 is the limit of a sequence $x_1, x_2, \ldots$ of points of X all distinct from x_0 and that

$$y_i \in F(x_i) \backslash F(x_0), \quad i \geq 1.$$

We need to prove that there is a point y_0 that is the norm limit of a subsequence of the sequence $y_1, y_2, y_3, \ldots$.

Since F, regarded as a map from X to $C_p(K)$, has the diagonal property, there is a point z_0 in $F(x_0)$ that is the pointwise limit of a subsequence, say $z_1, z_2, \ldots,$ of $y_1, y_2, y_3, \ldots$. Now

$$D = \{z_i : i \geq 0\}$$

is compact in $C_p(K)$. Further, D is bounded in $C(K)$, since it is contained in Z. By the results of Grothendieck, quoted above, it follows that D is weakly compact in $C(K)$ and that the weak and pointwise topologies coincide on D. Hence z_0 in $F(x_0)$ is the weak limit of $z_1, z_2, z_3, \ldots$.

Now z_0 in $F(x_0)$ is the norm limit of a sequence of finite rational convex combinations of the points $y_1, y_2, y_3, \ldots$, as required. $\square$

6.2 SELECTION THEOREMS

In this section we prove the main selection theorems stated in the introduction. We first prove three lemmas that provide the basis for the selection processes in the theorems. In each lemma, the function f provides a selector for the set-valued function

$$F_\epsilon = \{y : d(y, F(x)) < \epsilon\}.$$

Lemma 6.5 *Let X and Y be metric spaces. Let F be an upper semi-continuous set-valued map from X to Y, taking only nonempty values. Let y_0 be a given point of $F(X)$ and let $\epsilon > 0$ and $\delta > 0$ be given. Then there is a disjoint discretely σ-decomposable family $\mathcal{U}$ of $\mathcal{F}_\sigma$-sets covering X and a function f from X to Y satisfying the following conditions:*

(1) for each U in $\mathcal{U}$, diam $U < \delta$;

(2) $f(x) = y_0$, whenever $y_0 \in F(x)$;

(3) f is constant on each set U in $\mathcal{U}$ and takes a value in $F(U)$;

(4) for each x in X,

$$F(x) \cap B(f(x); \epsilon) \neq \emptyset.$$

Lemma 6.6   *Let X be a metric space and let Z be a convex subset of a Banach space Y. Let F be an upper semi-continuous set-valued map from X to Z with its weak topology, taking only nonempty values. Let y_0 be a given point in F(X) and let $\epsilon > 0$ and $\delta > 0$ be given. Then there is a disjoint discretely σ-decomposable family $\mathcal{U}$ of $\mathcal{F}_\sigma$-sets covering X and a function f from X to Z satisfying the conditions (1)–(4) of Lemma 6.5.*

Lemma 6.7   *Let X be a metric space and let K be a compact Hausdorff space. Let Z be a bounded convex subset of the space C(K) of continuous real-valued functions on K. Let F be a pointwise upper semi-continuous set-valued map from X to C(K), taking only nonempty values contained in Z. Let y_0 be a given point of F(z) and let $\epsilon > 0$ and $\delta > 0$ be given. Then there is a disjoint discretely σ-decomposable family $\mathcal{U}$ of $\mathcal{F}_\sigma$-sets covering X and a function f from X to Z satisfying the conditions (1)–(4) of Lemma 6.5.*

Proof of Lemma 6.5. Write

$$V_0 = \{x : y_0 \in F(x)\}.$$

Then $V_0 = F^{-1}(\{y_0\})$ is a closed subset of X. Since X is a metric space, we can choose, inductively, discretely σ-decomposable partitions $\mathcal{V}^{(n)}$ of $X \backslash V_0$ into nonempty $\mathcal{F}_\sigma$-sets, with

$$\operatorname{diam} V^{(n)} < \delta/n, \quad \text{for } V^{(n)} \in \mathcal{V}^{(n)},$$

and with $\mathcal{V}^{(n+1)}$ a refinement of $\mathcal{V}^{(n)}$ for $n \geq 1$. For each $V^{(n)}$ in $\mathcal{V}^{(n+1)}$ choose

$$x^{(n)} = x^{(n)}(V^{(n)}) \in V^{(n)}$$

and

$$y^{(n)} = y^{(n)}(V^{(n)}) \in F(x^{(n)}).$$

Define a function $g^{(n)} : X \rightarrow Y$, by taking

$$g^{(n)}(x) = y_0, \quad x \in V_0,$$

$$g^{(n)}(x) = y^{(n)}(V^{(n)}), \quad \text{if } x \in V^{(n)} \in \mathcal{V}^{(n)},$$

for each $n \geq 1$. Note that, for each x in $X \backslash V_0$, we have a unique sequence of sets $V^{(n)}$, $n \geq 1$, with

$$x \in V^{(n)} \in \mathcal{V}^{(n)}, \quad n \geq 1,$$

and corresponding sequences $x^{(n)}$, $n \geq 1$, and $y^{(n)}$, $n \geq 1$, with

$$x^{(n)} \in V^{(n)}, \quad n \geq 1,$$

$$y^{(n)} \in F(x^{(n)}), \quad n \geq 1,$$

$$y^{(n)} = g^{(n)}(x), \quad n \geq 1,$$

and

$$x^{(n)} \to x \quad \text{as } n \to \infty.$$

Perhaps

$$y^{(n)} = g^{(n)}(x) \in F(x)$$

for infinitely many $n \geq 1$. If not, we have

$$y^{(n)} \in F(x^{(n)}) \backslash F(x)$$

for all sufficiently large n. Hence, by Lemma 6.1, there will be a subsequence of the sequence $y^{(n)}$, $n \geq 1$, converging to some point of $F(x)$. Thus, in either case, there will be infinitely many n for which

$$F(x) \cap B\left(g^{(n)}(x); \frac{1}{2}\epsilon\right) = F(x) \cap B\left(y^{(n)}; \frac{1}{2}\epsilon\right) \neq \emptyset.$$

This shows that

$$\bigcup_{n=1}^{\infty} \left\{ x : F(x) \cap \overline{B}\left(g^{(n)}(x); \frac{1}{2}\epsilon\right) \neq \emptyset \right\} = X,$$

on remembering that

$$g^{(n)}(x) = y_0 \in F(x),$$

when $x \in V_0$.

Write

$$X^{(n)} = \left\{ x : F(x) \cap \overline{B}\left(g^{(n)}(x); \frac{1}{2}\epsilon\right) \neq \emptyset \right\},$$

$$\Xi^{(n)} = X^{(n)} \backslash \bigcup_{r=1}^{n-1} X^{(r)}$$

for $n \geq 1$. Then $\Xi^{(n)}$, $n \geq 1$, are disjoint sets with union X. Since $\mathcal{V}^{(n)}$ refines $\mathcal{V}^{(r)}$ for $1 \leq r \leq n-1$, the functions $g^{(r)}$, $1 \leq r \leq n$, are constant on each set $V^{(n)}$ in $\mathcal{V}^{(n)}$. Thus the sets

$$V^{(n)} \cap X^{(r)}, \quad 1 \leq r \leq n,$$

are relatively closed in $V^{(n)}$, and

$$V^{(n)} \cap \Xi^{(n)} = \left(V^{(n)} \cap X^{(n)} \right) \Big\backslash \left(\bigcup_{r=1}^{n-1} V^{(n)} \cap X^{(r)} \right)$$

is an $\mathcal{F}_\sigma$-set in X. Let $\mathcal{W}$ be the family of nonempty sets amongst the sets

$$V_0 \quad \text{and} \quad \Xi^{(n)} \cap V^{(n)} \quad \text{with} \quad n \geq 1 \quad \text{and} \quad V^{(n)} \in \mathcal{V}^{(n)}.$$

Then $\mathcal{W}$ is a discretely σ-decomposable partition of X into nonempty $\mathcal{F}_\sigma$-sets. Now, if $x \in X\backslash V_0$, there is a unique $W \in \mathcal{W}$, a unique $n \geq 1$, and a unique $V^{(n)}$ in $\mathcal{V}^{(n)}$ with

$$x \in W = \Xi^{(n)} \cap V^{(n)}.$$

Since $x \in X^{(n)}$, we have

$$F(x) \cap \overline{B}\left(g^{(n)}(x); \frac{1}{2}\epsilon \right) \neq \emptyset$$

and

$$g^{(n)}(x) = y^{(n)}(V^{(n)}),$$

since $x \in V^{(n)}$. Similarly, for all ξ in W,

$$F(\xi) \cap \overline{B}\left(g^{(n)}(\xi); \frac{1}{2}\epsilon \right) \neq \emptyset,$$

with

$$g^{(n)}(\xi) = y^n(V^{(n)}).$$

So we can choose $\xi(W)$ and $\eta(W)$ depending on W, but not on the position of x in W, with

$$\xi(W) \in W,$$

$$\eta(W) \in F(\xi(W)) \cap \overline{B}\left(y^{(n)}(V^{(n)}); \frac{1}{2}\epsilon \right).$$

This choice ensures that

$$d\left(\eta(W), y^{(n)}(V^{(n)}) \right) \leq \frac{1}{2}\epsilon,$$

so that

$$F(\xi) \cap B(\eta(W); \epsilon) \supset F(\xi) \cap \overline{B}\left(y^{(n)}(V^{(n)}); \frac{1}{2}\epsilon \right) \neq \emptyset$$

for $\xi \in W$. We can now define $f : X \rightarrow Y$ by taking

$$f(x) = y_0, \quad \text{if } x \in V_0,$$

$$f(x) = \eta(W), \quad \text{if } x \notin V_0, \text{ but } x \in W \in \mathcal{W}.$$

This ensures that: $f(x) = y_0$ whenever $y_0 \in F(x)$; f is constant on each set W in $\mathcal{W}$ and takes a value in $F(W)$; and

$$F(x) \cap B(f(x); \epsilon) \neq \emptyset$$

for each x in X. Further, diam $W < \delta$ for each W in $\mathcal{W}$, with the possible exception of V_0. We now need to replace the set V_0 by a union of sets of small diameter. Choose a disjoint discretely σ-decomposable family $\mathcal{T}$ of nonempty $\mathcal{F}_\sigma$-sets, each of diameter less than δ, and with union V_0. Take $\mathcal{U}$ to be the family

$$\mathcal{U} = \mathcal{T} \cup (\mathcal{W} \backslash \{V_0\}).$$

Then f and $\mathcal{U}$ satisfy our requirements. $\square$

Proof of Lemma 6.6. As in the proof of Lemma 6.5, define V_0, choose the families $\mathcal{V}^{(n)}$, the points

$$x^{(n)} = x^{(n)}(V^{(n)}), \quad y^{(n)} = y^{(n)}(V^{(n)})$$

for $V^{(n)} \in \mathcal{V}^{(n)}$, and define the functions $g^{(n)}$, but now as functions from X to Z, all for $n \geq 1$. Further note that for each x in $X \backslash V_0$ there are again unique sequences $V^{(n)}, x^{(n)}, y^{(n)}$ with

$$x \in V^{(n)} \in \mathcal{V}^{(n)},$$

$$x^{(n)} = x^{(n)}(V^{(n)}) \in V^{(n)},$$

$$y^{(n)} = y^{(n)}(V^{(n)}) \in F(x^{(n)}),$$

$$y^{(n)} = g^{(n)}(x)$$

for $n \geq 1$, and with

$$x^{(n)} \to x \quad \text{as } n \to \infty.$$

Perhaps

$$y^{(n)} = g^{(n)}(x) \in F(x)$$

for infinitely many $n \geq 1$. If not, we have

$$y^{(n)} \in F(x^{(n)}) \backslash F(x)$$

for all sufficiently large n. Hence, by Lemma 6.2, there will be a sequence of finite rational convex combinations of the points

$$y^{(n)} = g^{(n)}(x), \quad n \geq 1,$$

converging in norm to some point of $F(x)$. This ensures that, in either case, there will be a finite rational convex combination, say $\gamma(x)$, of the points $g^{(n)}(x)$, $n \geq 1$, with

$$F(x) \cap B\left(\gamma(x); \frac{1}{2}\epsilon\right) \neq \emptyset.$$

Let $h^{(m)}$, $m = 1, 2, \ldots$, be an enumeration of all the finite rational convex combinations of the functions $g^{(n)}$, $n \geq 1$. Then, for each x in X, there will be an $m \geq 1$ with

$$h^{(m)}(x) = \gamma(x)$$

and so

$$F(x) \cap B\left(h^{(m)}(x); \frac{1}{2}\epsilon\right) \neq \emptyset$$

for this m. For each $m \geq 1$, let

$$n_i = n_i(m), \quad 1 \leq i \leq k = k(m),$$

and let

$$1 \leq n_1 < n_2 < \cdots < n_k$$

be the integers n for which $g^{(n)}$ is used in forming the finite rational convex combination $h^{(m)}$. Let

$$N(m) = \max\{n_{k(r)}(r) : 1 \leq r \leq m\}.$$

Since

$$\mathcal{V}^{(N(m))} \text{ refines } \mathcal{V}^{(N)}, \quad \text{for } 1 \leq N \leq N(m),$$

the functions

$$g^{(N)}, \quad 1 \leq N \leq N(m),$$

and so also the functions

$$h^{(r)}, \quad 1 \leq r \leq m,$$

are constant on each member of $\mathcal{V}^{(N(m))}$. To simplify the notation write

$$\mathcal{W}^{(m)} = \mathcal{V}^{(N(m))}$$

and

$$z^{(m)}\left(\mathcal{W}^{(m)}\right) = y^{(N(m))}\left(\mathcal{V}^{(N(m))}\right)$$

for

$$W^{(m)} = V^{(N(m))},$$

with $W^{(m)}$ in $\mathcal{W}^{(m)} = \mathcal{V}^{(N(m))}$.

Write

$$X^{(m)} = \left\{ x : F(x) \cap \overline{B}\left(h^{(m)}(x); \frac{1}{2}\epsilon\right) \neq \emptyset \right\}$$

$$\Xi^{(m)} = X^{(m)} \Big\backslash \bigcup_{r=1}^{m-1} X^{(r)}$$

for $m \geq 1$. Then $\Xi^{(m)}$, $m \geq 1$, are disjoint sets with union X. As we have seen, the functions $h^{(r)}$, $1 \leq r \leq m$, are constant on each set $W^{(m)}$ of $\mathcal{W}^{(m)}$. Thus the sets

$$W^{(m)} \cap X^{(r)} = W^{(m)} \cap \left\{ x : F(x) \cap \overline{B}\left(h^{(r)}(x); \frac{1}{2}\epsilon\right) \neq \emptyset \right\}$$

$1 \leq r \leq m$ are relatively closed in $W^{(m)}$ and

$$W^{(m)} \cap \Xi^{(m)} = \left(W^{(m)} \cap X^{(m)}\right) \Big\backslash \left(\bigcup_{r=1}^{m-1} W^{(m)} \cap X^{(r)}\right).$$

is an $\mathcal{F}_\sigma$-set in X. Let $\mathcal{W}$ be the family of nonempty sets amongst the sets

$$V_0 \quad \text{and} \quad \Xi^{(m)} \cap W^{(m)}, \quad \text{with} \quad m \geq 1, \quad \text{and} \quad W^{(m)} \in \mathcal{W}^{(m)}.$$

Then $\mathcal{W}$ is a discretely σ-decomposable partition of X into $\mathcal{F}_\sigma$-sets. The rest of the proof can now be completed following the last three paragraphs of the proof of Lemma 6.5, making only obvious notational changes. $\qquad \square$

Proof of Lemma 6.7. The proof follows the proof of Lemma 6.6 with minor changes. It is necessary to work with the space $C_p(K)$ rather than with the space Y, and to use Lemma 6.4 in place of Lemma 6.2. One needs to note that, since F is pointwise upper semi-continuous to $C(K)$, taking its values in Z, F is also pointwise upper semi-continuous to Z. Further, each norm closed ball, such as $\overline{B}(g^{(n)}(x); \frac{1}{2}\epsilon)$ in Z is a closed set in the pointwise topology of Z. $\qquad \square$

Proof of Theorem 6.1. We suppose that X and Y are metric spaces and that F is an upper semi-continuous set-valued map from X to Y taking only nonempty values.

Take y_0 to be any point of $F(X)$ and write

$$f_0(x) = y_0, \quad \text{for } x \in X.$$

We define a sequence f_n, $n \geq 0$, of functions from X to Y and a sequence $\mathcal{U}_n$,

$n \geq 1$, of discretely σ-decomposable partitions of X into $\mathcal{F}_\sigma$-sets, with the following properties:

(1) $\mathcal{U}_n$ refines $\mathcal{U}_{n-1}$ for $n \geq 2$;

(2) when $n \geq 1$ and $f_{n-1}(x) \in F(x)$, then $f_n(x) = f_{n-1}(x)$;

(3) the function f_n is constant on each set U in $\mathcal{U}_n$ and takes a value in

$$F(U) \quad \text{when } n = 1$$

and in

$$F(U) \cap \overline{B}\left(f_{n-1}(x); \left(\frac{1}{2}\right)^n\right) \quad \text{when } n \geq 2;$$

(4) for each $n \geq 1$ and each x in X,

$$F(x) \cap \overline{B}\left(f_n(x); \left(\frac{1}{2}\right)^n\right) \neq \emptyset;$$

(5) for each $n \geq 1$ and each U in $\mathcal{U}_n$,

$$\operatorname{diam} U < \left(\frac{1}{2}\right)^n.$$

To start the construction we apply Lemma 6.5 with the chosen point y_0 and with $\epsilon = \delta = \frac{1}{2}$. This yields a function f_1 from X to Y and a discretely σ-decomposable partition $\mathcal{U}_1$ of X into $\mathcal{F}_\sigma$-sets satisfying the conditions (2), (3), (4) and (5) with $n = 1$.

Now suppose that $n \geq 1$ and that the functions f_r and the partitions $\mathcal{U}_r$ have been chosen for $1 \leq r \leq n$ satisfying our requirements. For U in $\mathcal{U}_n$ we consider set-valued function $F_U : U \to Y$, defined by

$$F_U(x) = F(x) \cap \overline{B}\left(f_n(x); \left(\frac{1}{2}\right)^n\right)$$

$$= F(x) \cap \overline{B}\left(y_n(U); \left(\frac{1}{2}\right)^n\right),$$

where $y_n(U)$ is the constant value taken by f_n on U. Since $\overline{B}(y_n(U); (\frac{1}{2})^n)$ is a fixed closed set, it follows from condition (4) that F_U is an upper semi-continuous map from U to Y taking only nonempty values. We now apply Lemma 6.5 with $X = U$, $F = F_U$, $y_0 = y_n(U)$, $\epsilon = \delta = (\frac{1}{2})^{n+1}$, to obtain a function $f_{n+1}^{(U)}$ and a discretely σ-decomposable partition $\mathcal{U}_{n+1}^{(U)}$ of U into $\mathcal{F}_\sigma$-sets satisfying the conditions (2)–(5) when we confine our attentions to U and to its partition $\mathcal{U}_{n+1}^{(U)}$ and replace n by $n + 1$.

Write

$$\mathcal{U}_{n+1} = \bigcup\{\mathcal{U}_{n+1}^{(U)} : U \in \mathcal{U}_n\}$$

and

$$f_{n+1}(x) = f_{n+1}^{(U)}(x) \quad \text{for } x \in U \in \mathcal{U}_n.$$

By Lemma 2.3 the family $\mathcal{U}_{n+1}$ is a discretely σ-decomposable partition of X into $\mathcal{F}_\sigma$-sets refining the family $\mathcal{U}_n$. It is now easy to verify that the conditions (1)–(5) hold with n replaced by $n+1$. In this way, the families f_n, $n \geq 0$, and $\mathcal{U}_n$, $n \geq 1$, can be constructed inductively satisfying our requirements.

We concentrate, for a time, on a fixed value of x. It may be that $f_N(x) \in F(x)$ for some $N \geq 1$. In this case, it follows by (2) that

$$f_n(x) = f_N(x) \in F(x)$$

for all $n \geq N$. Thus, in this case, $f_n(x)$ converges to a point of $F(x)$. Otherwise,

$$f_n(x) \notin F(x), \quad \text{for } n \geq 1.$$

By (3),

$$f_n(x) \in F(\xi_n)$$

for some ξ_n in the member U of $\mathcal{U}_n$ containing x. By (5), we have

$$\xi_n \in B\left(x, \left(\frac{1}{2}\right)^n\right)$$

for $n \geq 1$. Thus $\xi_n \to x$ as $n \to \infty$, and

$$\eta_n = f_n(x) \in F(\xi_n)\backslash F(x)$$

for $n \geq 1$. By Lemma 6.1, the sequence η_n, $n \geq 1$, has a subsequence converging to a point of $F(x)$. But condition (3) ensures that

$$\eta_n = f_n(x), \quad n \geq 1$$

is a Cauchy sequence. Hence, in this second case, $f_n(x)$ converges to a point of $F(x)$.

Write

$$f(x) = \lim_{n \to \infty} f_n(x)$$

for all x in X. By the results of the last paragraph, $f(x)$ is well defined and belongs to $F(x)$ for each x in X. Thus f is a selector for F. Further, f is the uniform limit of the functions f_n, each of these being constant on the sets of a discretely σ-decomposable family of $\mathcal{F}_\sigma$-sets. Thus each f_n and so also f is σ-discrete and of the first Borel class (see Theorem 2.1). $\square$

Proof of Theorem 6.2. We suppose that X is a metric space and Z is a convex subset of a Banach space Y. Suppose also that F is an upper semi-continuous set-valued map from X to Z, with the weak topology of Y, taking only non-empty values.

As in the first part of the proof of Theorem 6.1, but using Z in place of Y and Lemma 6.6 in place of Lemma 6.5, we define a sequence f_n, $n \geq 0$, of functions from X to Z and a sequence $\mathcal{U}_n$, $n \geq 1$, of discretely σ-decomposable partitions of X into $\mathcal{F}_\sigma$-sets, with the properties (1)–(5) of that proof. We repeat these properties, word for word, for the reader's convenience:

(1) $\mathcal{U}_n$ refines $\mathcal{U}_{n-1}$ for $n \geq 2$;

(2) when $n \geq 1$ and $f_{n-1}(x) \in F(x)$ then $f_n(x) = f_{n-1}(x)$;

(3) the function f_n is constant on each set U in $\mathcal{U}_n$ and takes a value in

$$F(U), \quad \text{when } n = 1,$$

and in

$$F(U) \cap \overline{B}\!\left(f_{n-1}(x); \left(\tfrac{1}{2}\right)^n\right), \quad \text{when } n \geq 2;$$

(4) for each $n \geq 1$ and each x in X,

$$F(x) \cap \overline{B}\!\left(f_n(x); \left(\tfrac{1}{2}\right)^n\right) \neq \emptyset;$$

(5) for each $n \geq 1$ and each U in $\mathcal{U}_n$,

$$\operatorname{diam} U < \left(\tfrac{1}{2}\right)^n.$$

We now concentrate, for a time, on a fixed value of x. It may be that $f_N(x) \in F(x)$ for some $N \geq 1$. In this case, it follows by (2) that

$$f_n(x) = f_N(x) \in F(x)$$

for all $n \geq N$. Thus, in this case, $f_n(x)$ converges to a point of $F(x)$. Otherwise,

$$f_n(x) \notin F(x), \quad \text{for } n \geq 1.$$

By (3),

$$f_n(x) \in F(\xi_n)$$

for some ξ_n in the member U of $\mathcal{U}_n$ containing x. By (5), we have

$$\xi_n \in B\!\left(x, \left(\tfrac{1}{2}\right)^n\right)$$

for $n \geq 1$. Thus $\xi_n \to x$ as $n \to \infty$, and

$$\eta_n = f_n(x) \in F(\xi_n) \backslash F(x)$$

for $n \geq 1$. The condition (3) ensures that the sequence

$$\eta_n = f_n(x), \quad n \geq 1,$$

is a Cauchy sequence in the Banach space Y, and so converges to some point, $f(x)$ say, in Y, but not as far as we know yet in Z. However, the set

$$S = \{f(x)\} \cup \{\eta_n : n \geq 1\}$$

is closed in Y so that $S \cap Z$ is relatively closed in Z, in both cases using the weak topology. Hence

$$F^{-1}(S \cap Z)$$

is closed in X, and containing the points ξ_n, $n \geq 1$, must also contain the point x. Thus

$$f(x) \in F(x) \subset Z$$

and in this case also

$$f(x) = \lim_{n \to \infty} f_n(x)$$

exists and belongs to Z.

Now f is a selector for F and, being the uniform limit of the functions f_n, each a function that is constant on the sets of a discretely σ-decomposable family of $\mathcal{F}_\sigma$-sets, is of the first Baire class as a map from X to Z with the norm topology (see Theorem 2.1). $\square$

Proof of Theorem 6.3. We suppose that X is a metric space, that K is a compact Hausdorff space and that Z is a convex subset of the Banach space $C(K)$ of continuous real-valued functions on K. Suppose that F is a pointwise upper semi-continuous set-valued map from X to Z, taking only nonempty values.

We first consider the case when Z is norm bounded in $C(K)$. In this case we follow the proof of Theorem 6.2, but using $C_p(K)$ for Y and Lemma 6.7 for Lemma 6.6. We obtain a selector $f : X \to Z$ for F that is of the first Baire class for the norm topology on Z.

Now consider the case of a general set Z. Write

$$Z_n = Z \cap (nB(K)),$$

where $B(K)$ is the closed unit ball of $C(K)$,

$$F_n(x) = F(x) \cap Z_n,$$

$$X_n = F^{-1}(Z_n).$$

Then

$$\bigcup_{n=1}^{\infty} X_n = X$$

and F_n is a pointwise upper semi-continuous set-valued map from X_n to Z_n, taking only nonempty values, for $n \geq 1$. By the result of the last paragraph there will be a selector $f_n : X_n \to Z_n$ for F_n that is of the first Baire class for the norm topology of Z_n. Now write

$$\Xi_n = X_n \setminus \left(\bigcup_{r=1}^{n-1} X_r \right)$$

and

$$f(x) = f_n(x) \quad \text{when } x \in \Xi_n.$$

This ensures that $f : X \to Z$ is a selector for F that is of the first Baire class.

Since Z is a convex set in $C(K)$, this selector is of the first Baire class, as required. $\quad \square$

6.3 A SELECTION THEOREM FOR LOWER SEMI-CONTINUOUS SET-VALUED MAPS

In this section we give a simple proof of Theorem 6.4.

Proof of Theorem 6.4. Let X and Y be metric spaces and let F be a lower semi-continuous set-valued map from X to Y taking only nonempty values that are complete in the metric on Y. Choose a sequence $\mathcal{V}_n$, $n \geq 1$, of open covers of Y with

$$\text{diam } V < 2^{-n}, \quad \text{when } V \in \mathcal{V}_n,$$

and arrange that

$$\mathcal{V}_{n+1} \text{ refines } \mathcal{V}_n,$$

when $n \geq 1$. For each $n \geq 1$, let $\mathcal{U}_n$ be the corresponding family

$$\{ U = F^{-1}(V) : V \in \mathcal{V}_n \}.$$

Since F is lower semi-continuous, $\mathcal{U}_n$ is an open cover of X and $\mathcal{U}_{n+1}$ refines $\mathcal{U}_n$ for $n \geq 1$.

By Lemma 2.1, we can choose a discretely σ-decomposable partition $\mathcal{W}_1$ of X into $\mathcal{F}_\sigma$-sets, with $\mathcal{W}_1$ refining $\mathcal{U}_1$. By choosing a corresponding partition for $\mathcal{U}_2$, and then taking its intersection with $\mathcal{W}_1$ we obtain a discretely σ-decomposable partition $\mathcal{W}_2$ of X into $\mathcal{F}_\sigma$-sets, with $\mathcal{W}_2$ refining both $\mathcal{U}_2$ and $\mathcal{W}_1$. Proceeding inductively we obtain a sequence $\mathcal{W}_n$, $n \geq 1$, of σ-*decomposable* partitions of X into $\mathcal{F}_\sigma$-sets with

$$\mathcal{W}_n \text{ refines } \mathcal{U}_n,$$

$$\mathcal{W}_{n+1} \text{ refines } \mathcal{W}_n,$$

for $n \geq 1$.

For each $n \geq 1$ and each nonempty W_n in $\mathcal{W}_n$ choose a point $\eta_n(W_n)$ in V_n, where U_n is the unique set in $\mathcal{U}_n$ that contains W_n and V_n is the unique set in $\mathcal{V}_n$ with $U_n = F^{-1}(V_n)$. Define functions f_n and φ_n from X to Y by taking

$$f_n(x) = \eta_n(W_n),$$

when

$$x \in W_n \in \mathcal{W}_n,$$

and choosing

$$\varphi_n(x) \in F(x) \cap V_n,$$

when

$$x \in W_n \subset U_n = F^{-1}(V_n),$$

with $W_n \in \mathcal{W}_n$, $U_n \in \mathcal{U}_n$ and $V_n \in \mathcal{V}_n$ uniquely determined by this condition. Note that the choice of $\varphi_n(x)$ is always possible as the presence of x in $F^{-1}(V_n)$ ensures that $F(x) \cap V_n \neq \emptyset$.

Consider any x in X and any $n \geq 1$. Then

$$x \in W_{n+1} \subset W_n$$

for unique sets $W_n \in \mathcal{W}_n$ and $W_{n+1} \in \mathcal{W}_{n+1}$. Further $W_n \subset U_n$, $W_{n+1} \subset U_{n+1}$ for unique sets U_n, U_{n+1} in $\mathcal{U}_n$ and $\mathcal{U}_{n+1}$. Since

$$x \in U_n, \quad x \in U_{n+1},$$

we must have $U_{n+1} \subset U_n$. Thus the points

$$\varphi_n(x), \quad \varphi_{n+1}(x), \quad f_n(x), \quad f_{n+1}(x)$$

all lie in the unique set V_n in $\mathcal{V}_n$ with $F^{-1}(V_n) = U_n$, and

$$d(\varphi_{n+1}(x), \varphi_n(x)) < 2^{-n},$$

$$d(f_n(x), \varphi_n(x)) < 2^{-n}.$$

Thus the sequence $\varphi_n(x)$ is a Cauchy sequence lying in the complete set $F(x)$. Hence the formula

$$f(x) = \lim_{n \to \infty} \varphi_n(x)$$

defines a point of $F(x)$ and so f is a selector for F. Further we obtain

$$d(f_n(x), f(x)) \leq 2^{-n} + d(\varphi_n(x), f(x))$$

$$\leq 2^{-n+2},$$

so that the sequence of functions f_n converges uniformly to f. Since each function f_n is constant on the sets of a discretely σ-decomposable partition of X into $\mathcal{F}_\sigma$-sets it follows by Theorem 2.1 that f is a σ-discrete function of the first Borel class. $\square$

6.4 EXAMPLE

In this section, following a suggestion of Valdivia, we construct an example.

Example   *There is a complete metric space X, a dual Banach space Y^* that is weak* σ-fragmented using weak* closed sets and an upper semi-continuous set-valued map F from X to Y^* with its weak* topology, taking only nonempty norm complete values in Y^* and having no selector f that is of the first Baire class as a map from X to Y^* with its norm topology.*

Construction   Let ω_1 be the first uncountable ordinal and let $[0, \omega_1]$ be the space of all ordinals γ with $0 \le \gamma \le \omega_1$ with the order topology. We take Y^* to be the dual space $C^*([0, \omega_1])$ of the space $C([0, \omega_1])$ of continuous real-valued functions on $[0, \omega_1]$ with the supremum norm. Since the space $[0, \omega_1]$ is scattered, the space $C([0, \omega_1])$ is an Asplund space (see the discussion on Asplund spaces in Chapter 7). Hence $C^*([0, \omega_1])$ has the Radon–Nikodým property and so $Y^* = C^*([0, \omega_1])$ with its weak* topology is σ-fragmented using weak* closed sets as required.

We take X to be the subspace of $C([0, \omega_1])$ of all nonnegative functions x on $[0, \omega_1]$ with

$$x(\omega_1) = \|x\|_\infty = 1$$

and use the supremum norm on $C([0, \omega_1])$ to define the metric on X. Clearly X is a complete metric space, as required.

We remark that $C^*([0, \omega_1])$ is the space of all measures μ on $[0, \omega_1]$. Since ω_1 is not a measureable cardinal, all these measures μ are atomic, and the norm on $C^*([0, \omega_1])$ is given by

$$\|\mu\|_1 = \sum \{|\mu(\gamma)| : 0 \le \gamma \le \omega_1\}.$$

We now define a set-valued map F from X to Y^*. We first take $\Gamma(x)$ to be the set

$$\Gamma(x) = \{\gamma : 0 \le \gamma \le \omega_1 \text{ and } x(\gamma) = 1\}$$

for each x in X. Note that $\omega_1 \in \Gamma(x)$ for each $x \in X$. For each x in X, take $F(x)$ to be the set

$$\{\mu \in Y^* : \|\mu\|_1 = 1, \ \mu(\gamma) \ge 0, \text{ for } 0 \le \gamma \le \omega_1$$

$$\text{and } \mu(\gamma) = 0 \text{ for } \gamma \notin \Gamma(x), \text{ and } \mu(\omega_1) = 0\}.$$

It is clear that F takes only norm complete values in Y^*. For each x in X, $x(\omega_1) = 1$. Since ω_1 has uncountable cofinality we have some δ with $0 < \delta < \omega_1$ and

$$x(\gamma) = 1, \quad \text{for } \delta \leq \gamma \leq \omega_1.$$

This ensures that $\delta \in \Gamma(x)$ so that the Dirac measure on δ belongs to $F(x)$. Thus F takes only nonempty norm complete values, as required. We prove that the set-valued map

$$F : X \longrightarrow (Y^*, \text{weak}^*)$$

is upper semi-continuous. Consider any x_0 in X, and any weak* open set G with

$$F(x_0) \subset G.$$

We need to prove that for some $\epsilon > 0$,

$$F(\{x \in X : \|x - x_0\|_\infty < \epsilon\}) \subset G.$$

Suppose that this is not true. Then we can choose x_i, μ_i, $i = 1, 2, \ldots$ with $x_i \in X$, $\|x_i - x_0\|_\infty < 1/i$, and $\mu_i \in F(x_i)\backslash G$, for $i \geq 1$. Then

$$\mu_i \in K = \{\mu : \|\mu\|_1 \leq 1\} \backslash G$$

for $i \geq 1$, and K is weak* compact. By the weak* compactness of K, we can choose κ in K with the property that each weak* neighborhood of κ contains μ_i for infinitely many values of i. We obtain a contradiction by studying all the possible forms for κ.

Case (a) Suppose that $\kappa(\gamma) < 0$ for some γ in $[0, \omega_1]$. Since

$$\kappa([0, \gamma]) = \sum \{\kappa(\eta) : 0 \leq \eta \leq \gamma\} \leq \|\kappa\|_1 \leq 1$$

for some δ, either 0 or a successor ordinal, we have

$$0 \leq \delta \leq \gamma \quad \text{and} \quad \kappa([\delta, \gamma]) < 0.$$

Define y_0 in $Y = C([0, \omega_1])$ by taking

$$y_0(\eta) = \begin{cases} 1, & \text{if } \eta \in [\delta, \gamma], \\ 0, & \text{otherwise.} \end{cases}$$

Then

$$\langle y_0, \kappa \rangle < 0,$$

but

$$\langle y_0, \mu \rangle \geq 0$$

for all μ in $F(X)$. Thus

$$N = \{\mu \in Y^* : \langle y_0, \mu \rangle < 0\}$$

is a weak* neighborhood of κ containing none of the points μ_i, $i \geq 1$.

Case (b)   Suppose that $\|\kappa\|_1 < 1$. Consider the point y in Y with

$$y(\gamma) = 1, \quad \text{for} \quad 0 \leq \gamma \leq \omega_1.$$

Then the set

$$N = \left\{ \mu : \langle y, \mu \rangle < \frac{1}{2} + \frac{1}{2} \|\kappa\|_1 \right\}$$

is a weak* neighborhood of κ containing none of the points μ_i, $i \geq 1$.

Case (c)   Suppose that $\kappa(\gamma) > 0$ for some γ with $0 \leq \gamma \leq \omega_1$ and $\gamma \notin \Gamma(x_0)$. Since x_0 is continuous on $[0, \omega_1]$ the set $\Gamma(x_0)$ is closed in $[0, \omega_1]$. So we can choose δ with $0 \leq \delta \leq \gamma$, δ either 0 or a successor ordinal,

$$[\delta, \gamma] \cap \Gamma(x_0) = \emptyset,$$

and

$$\kappa([\delta, \gamma]) > 0.$$

Define y in Y by taking

$$y(\eta) = \begin{cases} 1, & \text{if } \delta \leq \eta \leq \gamma, \\ 0, & \text{otherwise.} \end{cases}$$

Then

$$N = \{\mu : \langle y_0, \mu \rangle > 0\}$$

is a weak* neighborhood of κ. Since $[\delta, \gamma]$ is compact in $[0, \omega_1]$ and x_0 is continuous with

$$x_0(\eta) < 1, \quad \text{for} \quad \delta \leq \eta \leq \gamma,$$

we have

$$\sup\{x_0(\eta) : \delta \leq \eta \leq \gamma\} < 1.$$

Thus

$$\epsilon = 1 - \sup\{x_0(\eta) : \delta \leq \eta \leq \gamma\}$$

is positive. Hence, for each x in X with

$$\|x - x_0\|_\infty < \epsilon,$$

we have

$$x(\eta) < 1, \quad \text{for} \quad \delta \leq \eta \leq \gamma,$$

so that

$$\Gamma(x) \cap [\delta, \gamma] = \emptyset$$

and

$$\langle y, \mu \rangle = 0$$

for all μ in $F(x)$ with $\|x - x_0\|_\infty < \epsilon$. Thus N contains none of the points μ_i with $i \geq 1/\epsilon$.

Case (d) Suppose that $\kappa(\omega_1) > 0$. For each $i \geq 1$, $\mu_i(\omega_1) = 0$, and since ω_i is of uncountable cofinality, we can choose $\delta < \omega_1$ with

$$\mu_i(\eta) = 0, \quad \text{for} \quad \delta_i \leq \eta \leq \omega_1.$$

Now we can choose a successor ordinal δ with

$$\delta_i \leq \delta \leq \omega_1, \text{ for } i \geq 1.$$

Define y in Y by taking

$$y(\eta) = \begin{cases} 1, & \text{if } \delta \leq \eta \leq \omega_1, \\ 0, & \text{otherwise.} \end{cases}$$

Then

$$N = \left\{ \mu : \langle y, \mu \rangle > \frac{1}{2}\kappa(\omega_1) \right\}$$

is a weak* neighborhood of κ containing none of the points μ_i, $i \geq 1$.

In each of the cases (a) to (d) we reach a contradiction. Thus κ must satisfy the conditions

$$\kappa(\gamma) \geq 0, \quad \text{for } 0 \leq \gamma \leq \omega_1,$$

$$\|\kappa\|_1 \geq 1,$$

$$\kappa(\gamma) = 0, \quad \text{when } \gamma /\in \Gamma(x_0),$$

$$\text{and } \kappa(\omega_1) = 0.$$

Since we also have $\|\kappa\|_1 \leq 1$, from $\kappa \in K$, we conclude that $\kappa \in F(x_0)$. Since $F(x_0) \subset G$, this contradicts the choice of κ. Hence F is upper semi-continuous as a map to (Y^*, weak^*).

We now consider any selector f for F. We study f in a neighborhood of any point x_0 of X. Write $\mu_0 = f(x_0)$. Then $\mu_0(\gamma)$ is positive for countably many γ and $\mu_0(\omega_1) = 0$. Hence we can choose an ordinal δ with $0 < \delta < \omega_1$ and

$$\mu_0(\gamma) = 0, \quad \text{for } \delta \leq \gamma \leq \omega_1.$$

Take ζ to be the function in X defined by

$$\zeta(\gamma) = \begin{cases} 1, & \text{for} \quad 0 \le \gamma \le \delta, \\ 0, & \text{for} \quad \delta + 1 \le \gamma \le \omega_1. \end{cases}$$

For each $\epsilon > 0$, write

$$x_\epsilon(\gamma) = \max\{0, x_0(\gamma) - \epsilon\zeta(\gamma)\}$$

for $0 \le \gamma \le \omega_1$. Then x_ϵ belongs to X with

$$\|x_0 - x_\epsilon\|_\infty \le \epsilon$$

and

$$\Gamma(x_\epsilon) \subset [\delta + 1, \omega_1].$$

Now $\mu_\epsilon = f(x_\epsilon)$ is concentrated on $[\delta + 1, \omega_1]$, while $\mu_0 = f(x_0)$ is concentrated on $[0, \delta]$. Thus

$$\|f(x_0) - f(x_\epsilon)\|_1 = \|\mu_0 - \mu_\epsilon\|_1 = \|\mu_0\|_1 + \|\mu_\epsilon\|_1 = 2$$

and f is not norm continuous at x_0. If f were a function of the first Baire class from X to (Y^*, norm), then the set of points of norm discontinuity of f would be of the first category in X (see, e.g., [47, pp. 397–398]). Since X is complete, f must have a point of norm continuity. This contradiction shows that F has no selector that is of the first Baire class as a map from X to (Y^*, norm).

6.5 REMARKS

1. In the proof of Theorem 6.1, it would be possible to replace the condition:

    (a) F is upper semi-continuous;

    by the condition:

    (b) $F^{-1}(C)$ is closed in X whenever C is a closed countable set in Y.

    However the gain in generality would be spurious since the condition (b) implies the condition (a).

2. In Theorem 6.2 it would be possible, by making minor modifications to the proof, to replace the condition:

    (a) F is weakly upper semi-continuous;

    by the condition

    (b) $F^{-1}(C)$ is closed in X whenever C is convex, norm closed and norm separable in Z.

3. If one drops the condition that Z be convex in $C(K)$ in Theorem 6.3, one

still obtains a selector that is σ-discrete and of the first Borel class. This is clear from the proof that we have given.

4. Theorem 6.4 is not true if F is allowed to take nonempty but otherwise arbitrary values. Indeed, the result fails in a case when X and Y are the unit interval and F is a lower semi-continuous set-valued map taking only nonempty countable values.

5. We give a very simple example to illustrate the nature of the diagonal lemmas. Consider the set-valued map F from $\mathbb{R}$ to $\mathbb{R}^2$ that maps the points ξ of $\mathbb{R}$ to the discs

$$F(\xi) = \{(y, z) : y^2 + z^2 \leq (\rho(\xi))^2\}$$

in $\mathbb{R}^2$, where

$$\rho(\xi) = 1,$$

if ξ is not of the form $1/m$ with m a positive integer, while

$$\rho(\xi) = 1 + (1/m),$$

when ξ is of this form $1/m$ with m a positive integer. It is easy to verify that F is upper semi-continuous as a map from $\mathbb{R}$ to $\mathbb{R}^2$. Now suppose that $G(F)$ is the graph of F in $\mathbb{R}^3$, and that $(x(n), y(n), z(n))$, $n \geq 1$, is any "diagonal" sequence of points in $G(F)$ with $x(n)$ converging to some ξ in $\mathbb{R}$ and with

$$x(n) \neq \xi,$$

$$(y(n), z(n)) \in F(x(n)) \setminus F(\xi)$$

for $n \geq 1$. Then

$$F(x(n)) \setminus F(\xi) = \emptyset,$$

unless $x(n)$ is of the form $1/m(n)$, for some positive integer $m(n)$. Thus $x(n)$ is necessarily of this form, for each $n \geq 1$. Further, as $x(n)$ converges to ξ, while $x(n)$ does not assume the value ξ, we must have $\xi = 0$ and $m(n) \to \infty$ as $n \to \infty$. Now the point $(y(n), z(n))$ lies in the annular region

$$1 < y^2 + z^2 \leq 1 + (1/m(n))^2.$$

Thus the sequence $(y(n), z(n))$ has a subsequence converging to a point on the circle

$$y^2 + z^2 = 1$$

contained in the disc $F(0)$.

Chapter 7

Further applications

In Chapter 5 we have given some applications of the existence theorems for selectors of the first Baire class, giving detailed proofs based on well known basic results in general topology and Banach space theory. In this chapter we give some further applications, but we only give some of the details of the proofs and have to quote some difficult results without proofs, since the proofs would take us too far from our main themes.

Our results are centered around the theory of Asplund spaces. We start by discussing some of the work done on these spaces.

E. Asplund [1] introduced "strong differentiability spaces" defining them to be those real Banach spaces where every continuous convex function defined on a convex subset is Fréchet differentiable on a $\mathcal{G}_\delta$-subset of its domain. He obtained properties of the dual of such a Banach space and established the following result.

Theorem (Asplund) *If a real Banach space admits an equivalent norm whose dual norm is locally uniformly convex, then the original space is a strong differentiability space.*

After Asplund's untimely death, these strong differentiability spaces were renamed Asplund spaces by Namioka and Phelps in a paper [61] that developed the theory of these spaces. They prove that *a Banach space X is an Asplund space if and only if the dual space X^* has the following property:*

> *every nonempty bounded subset of X^* has nonempty relatively weak* open subsets of arbitrarily small diameter,*

that is, in our terminology,

> *every bounded subset of X^* is weak* fragmented by the norm.*

They show that the dual X^* of an Asplund space has the Radon–Nikodým property, so that by a result of Stegall [75] *every separable subspace of X has a separable dual*. They also prove that *every closed linear subspace of an Asplund space is an Asplund space*.

We note that in Theorem 5.2 we have used the existence of a Baire first class selector to give a new proof of the result of Namioka and Phelps that X is an

Asplund space if each bounded nonempty subset of X^* is weak* fragmented by the norm. The result of Asplund that we have quoted leaves open two questions.

> *If X is an Asplund space does X have an equivalent norm whose dual norm is locally uniformly convex?*
>
> *If X is an Asplund space does X^* have an equivalent norm that is locally uniformly convex?*

The answer to the first of these questions is "No". In Theorem 18 of their paper [61], Namioka and Phelps show that, *when K is a nonempty compact Hausdorff space, then C(K) is an Asplund space if and only if K is scattered.* In particular, if Ω is any ordinal, the ordinal interval $[0,\Omega]$ with its interval topology is a scattered compact Hausdorff space, and $C([0, \Omega])$ is an Asplund space. However, Talagrand [80] has shown that when Ω is uncountable $C([0, \Omega])$ has no equivalent norm with a strictly convex dual norm.

Thus, as we have said, the answer to the first question is "No". Neverthe less, Fabian and Godefroy [10] have recently shown that the answer to the second question is "Yes". We give an outline of their proof of their theorem in Sections 7.1 and 7.2 (see Theorem 7.2 in Section 7.2).

We note that the theorem of Namioka and Phelps, that *in the dual X^* of an Asplund space X every bounded subset is weak* fragmented by the norm*, is a converse to Theorem 5.2. Does the somewhat similar Theorem 5.4 have a converse? Recall that Theorem 5.4 tells us that *if K^* is a nonempty convex weak* compact set in the dual X^* of a Banach space, and K^* is weak* fragmented by the norm of X^*, then the attainment map from X to K^* has a selector that is of the first Baire class from $(X, norm)$ to $(X^*, norm)$.*

In Section 7.3 we prove the following partial converse to this result.

Theorem 7.1 *Let K^* be a nonempty weak* compact set in the dual X^* of a Banach space X. Suppose that the attainment map F from X to K^* has a selector f that is of the first Baire class as a map from $(X \ norm)$ to $(X^*, norm)$. Provided X contains no isomorphic copy of $\ell_1(\mathbb{N})$, the set K^* is weak* fragmented by the norm in X^*.*

The problem of finding a converse to Theorem 5.4 was discussed by Jayne, Orihuela, Pallarés and Vera [41]. We follow the proof of their Theorem 26 closely, but not too closely, since they claimed to prove the above theorem without the proviso that X contains no isomorphic copy of $\ell_1(\mathbb{N})$, while we need this proviso, and give Example 7.1 to show that some such proviso is necessary.

7.1 BOUNDARY LEMMAS

A subset S^* of a set K^* in the dual X^* of a Banach space X is said to be a *boundary* for K^* if, for each x in X, there is a point s^* in S^* with

$$\langle x, s^* \rangle = \sup\{\langle x, k^* \rangle : k^* \in K^*\}.$$

We prove two lemmas on boundaries in dual Banach spaces. The first is closely related to Theorem I.2 of Godefroy [15], and uses ideas from his proof, but avoids use of Simons' Lemma [72]. The second boundary lemma is tacitly included within the second step of the proof of Theorem 1 of Fabian and Godefroy [10]. We follow their proof closely.

We now state our two boundary lemmas.

Lemma 7.1 *Let X be a separable Banach space that contains no isomorphic copy of $\ell_1(\mathbb{N})$. Let K^* be a nonempty convex set in X^*, and let S^* be a boundary for K^*. Then K^* is contained in the norm closure of the linear span of S^*.*

Lemma 7.2 *Let X be a separable Banach space that contains no isomorphic copy of $\ell_1(\mathbb{N})$. Let K^* be a nonempty weak* compact convex set in X^*. Suppose that the attainment map F from X to K^* has a selector f that is of the first Baire class as a map from (X, norm) to (X^*, norm). Then $f(x)$ is a separable boundary for K^* and K^* is contained in a separable subspace of X^*.*

Proof of Lemma 7.1. Since S^* is a boundary for K^*, for each point $x \neq 0$ in X, there is a point s^* in S^* with

$$\langle x, s^* \rangle = \sup\{\langle x, x^* \rangle : x^* \in K^*\}.$$

Suppose that there is a point $k^* \neq 0$ in K^* that is not in $\overline{\text{sp}}S^*$, the norm closed linear span of S^*. Then, by the Hahn–Banach theorem, there is a linear functional j^{**} on X^* with

$$\langle \ell^*, j^{**} \rangle = 0 \text{ for all } \ell^* \text{ in } \overline{\text{sp}}S^*$$

and

$$\langle k^*, j^{**} \rangle = \|k^*\|, \quad \|j^{**}\| = 1.$$

We use a result of Odell and Rosenthall [64]. Since X is separable and there is no isomorphic copy of $\ell_1(\mathbb{N})$ in X, *each element of X^{**} can be obtained as the weak* limit in X^{**} of a sequence of points in X.* Applying this result to the point j^{**} chosen above, there will be a sequence $f_n, n \geq 1$, of X weak* convergent to j^{**} in X^{**}. In particular,

$$\lim_{n \to \infty} \langle k^*, f_n \rangle = \langle k^*, j^{**} \rangle = \|k^*\| \neq 0.$$

Thus, by omitting a finite set of points from the sequence, we may suppose that

$$\langle f_n, k^* \rangle > \frac{1}{2} \|k^*\|, \quad \text{for } n \geq 1.$$

This ensures that

$$\|f_n\| \geq \frac{1}{2}, \quad \text{for } n \geq 1.$$

Now, for each s^* in S^*, the weak* convergence to j^{**} ensures that

$$\lim_{n \to \infty} \langle f_n, s^* \rangle = \langle j^{**}, s^* \rangle = 0.$$

However, as we have seen

$$\lim_{n \to \infty} \langle f_n, k^* \rangle = \langle j^{**}, k^* \rangle = \|k^*\| \neq 0.$$

This proves the lemma. □

Proof of Lemma 7.2. Recall that the attainment map $F : X \to K^*$ is defined by

$$F(x) = \{x^* \in K^* : \langle x, x^* \rangle = \sup\{\langle x, y^* \rangle : y^* \in K^*\}\}.$$

This map has, by a hypothesis of the lemma, a selector f of the first Baire class as a map from (X, norm) to (X^*, norm). By this choice of f, for each x in X,

$$\langle x, f(x) \rangle = \sup\{\langle x, k^* \rangle : k^* \in K^*\}.$$

So for each x in X and each k^* in K^*,

$$\langle x, k^* \rangle \leq \langle x, f(x) \rangle \quad \text{and} \quad f(x) \in K^*.$$

This means that the set $f(X) = \{f(x) : x \in X\}$ is a boundary for K^* in X^*.

We next show that $f(X)$ has a countable norm dense set. Since f is of the first Baire class, we can write

$$f = \lim_{p \to \infty} f_p,$$

the limit being taken pointwise, and each function $f_p : X \to X^*$ being continuous from (X, norm) to (X^*, norm). Let $x_q, q \geq 1$, be a countable dense set in X. For each p, q we can choose a point $d^*(p, q)$ with $d^*(p, q) \in f(X)$ and

$$\left\| d^*(p, q) - f_p(x_q) \right\| \leq 2 \inf\{\left\| k^* - f_p(x_q) \right\| : k^* \in f(X)\}.$$

Now suppose that $\epsilon > 0$ is given and y^*, equal to $f(x)$, say, is chosen in $f(X)$. We can choose p so large that

$$\left\| f_p(x) - f(x) \right\| < \frac{1}{6} \epsilon.$$

Since f_p is continuous and $x_q, q \geq 1$, is dense in X, we can choose $q \geq 1$, so that

$$\left\|f_p(x_q) - f_p(x)\right\| < \frac{1}{6}\epsilon.$$

This ensures that

$$\left\|f_p(x) - f(x)\right\| < \frac{1}{3}\epsilon \quad \text{and} \quad f(x) \in f(X).$$

By the choice of $d^*(p, q)$ we have

$$\left\|d^*(p, q) - f_p(x_q)\right\| \leq 2\left\|f(x) - f_p(x_q)\right\| < \frac{2}{3}\epsilon.$$

Hence

$$\left\|d^*(p, q) - f(x)\right\| < \frac{2}{3}\epsilon + \frac{1}{3}\epsilon = \epsilon.$$

Thus the countable set

$$D_o = \{d(p, q) : p \geq 1, \ q \geq 1\}$$

is dense in $f(X)$, as required.

Now the countable set D_1 of all finite rational linear combinations of the points of D_0 is dense in the norm closed linear span L of $f(X)$. Since $f(X)$ is a boundary of K^*, Lemma 7.1 applies and shows that K^* is contained in the separable subspace L of X^*. $\square$

7.2 DUALS OF ASPLUND SPACES

In this section we give an outline of the way that Fabian and Godefroy [10] use a selection result as the first step of their theorem.

Theorem 7.2 (Fabian and Godefroy) *The dual of an Asplund space has a locally uniformly convex norm that is equivalent to the original norm.*

As we have already remarked, some Asplund spaces have no norms equivalent to their original norms whose dual norms are strictly convex.

We recall the definition of a locally uniformly convex norm on a Banach space. *A norm $\|\cdot\|$ on a Banach space X is said to be locally uniformly convex if, whenever a point x and a sequence $\{x_n\}$ of points in X satisfy the conditions*

$$\|x_n\| \to \|x\| \quad \text{and} \quad \left\|\frac{1}{2}x + \frac{1}{2}x_n\right\| \to \|x\| \text{ as } n \to \infty$$

then $\|x_n - x\| \to 0$ as $n \to \infty$. This condition implies that, near a point x on the unit sphere, the unit ball is uniformly strictly convex, the uniformity being with respect to the different directions of approach to x along the unit sphere.

Step 1. Consider the attainment map $F : X \to B^*$ from X to the unit ball B^*

of X^*. By the result of Namioka and Phelps quoted in the introductory section of this chapter, each bounded subset of X^* is weak* fragmented by the norm. Applying Theorem 5.4, we obtain a selector d for F that is of the first Baire class as a map from (X, norm) to (B^*, norm). Hence we can choose a sequence $\{d_n\}$ of continuous functions from (X, norm) to (B^*, norm) with d_n converging pointwise to d as $n \to \infty$. We use D to denote the set-valued map from X to B^*, defined by

$$D(x) = \{d_1(x), \ d_2(x), \ldots\}.$$

Note that since d is a selector for the attainment map, $d(X)$ is a boundary for B^*.

Step 2. Let V be a linear subspace of X. In this step we show that when V is a separable subspace of X, the closed linear span

$$\overline{\text{sp}}\{x^*|_V : x^* \in D(V)\}$$

coincides with the dual space V^* of V.

Note that when V is a linear subspace of X, any linear functional on V can be extended, without increase of norm, to yield a linear functional on X. So the linear functionals on V can be obtained (in general in many ways) as the restrictions to V of the linear functionals of X. Thus, we can (and we do) regard the dual space V^* of V as those distinct functionals $x^*|_V$ obtained by restricting an x^* in X^* to V; the norm in the dual V^* being

$$\|x^*|_V\| = \sup\{|x^*(v)| : v \in V, \ \|v\| = 1\}.$$

Since $d(X)$ is a boundary of B^* it is easy to see that

$$d|_V = \overline{\text{sp}}\{x^*|_V : x^* \in d(V)\}.$$

Since for each v in V,

$$d_n(v) \to d(v)$$

as $n \to \infty$. We also have

$$V^* = \overline{\text{sp}}\{x^*|_V : x^* \in D(V)\}.$$

Step 3. This next step shows that for any subspace V of X we still have

$$\overline{\text{sp}}\{x^*|_V : x^* \in D(V)\} = V^*.$$

The idea is to use a reduction of the general case to the separable case. We fix our attention on a particular point, say f, in the dual space V^* and aim to choose a suitable separable subspace Y of V. The point $f|_Y$ is automatically in Y^*. Since Y is separable, Step 2 applies and yields

$$f|_Y \in \overline{\text{sp}}\{x^*|_Y : x^* \in D(Y)\}.$$

The space Y has to be chosen in a very special way to enable the completion of the proof.

We shall obtain Y as the norm closure of

$$W = \bigcup_{n=1}^{\infty} W_n,$$

where $W_1, W_2, \ldots$ is an increasing sequence of separable subspaces of V, with dual spaces $W_1^*, W_2^*, \ldots$ passing ever closer to f.

We start by taking W_1 to be any one-dimensional subspace of V. When $n \geq 1$, and W_n has been chosen as a separable subspace of V, we choose a countable sequence $\{z_j^{(n)}\}$ dense in W_n. Since the functions $d_i, i \geq 1$, are continuous, the points

$$d_i(z_j^{(n)}), \quad i \geq 1, \quad j \geq 1,$$

will be dense in $D(W_n)$. We take $\mathcal{D}_n$ to be the set of all finite rational linear combinations of the points $d_i(z_j^{(n)}), i \geq 1, j \geq 1$. In the case when $n \geq 2$ and $W_{n-1} \subset W_n$, we choose the sequence $(z_j^{(n)})$ to include all the points of the sequence $(z_j^{(n-1)})$. Clearly $\mathcal{D}_n$ is a countable subset of X^*.

Once $\mathcal{D}_n$ has been chosen for some $n \geq 1$, for each d in $\mathcal{D}_n$ we note that

$$\|f|_V - d|_V\|_{V^*} = \sup\{\langle f - d, v \rangle : v \in V \text{ and } \|v\| = 1\}.$$

Thus we can choose a $v(n, d)$ in V with $\|v(n, d)\| = 1$ and

$$\langle f - d, v(n, d) \rangle \geq \|f|_V - d|_V\|_{V^*} - (1/n).$$

We use $\mathcal{V}_{n+1}$ to denote the set of all points $v(n, d)$ chosen in this way. This set $\mathcal{V}_{n+1}$ is a countable subset of V. We take

$$W_{n+1} = \overline{\mathrm{sp}}\{W_n \cup V_{n+1}\}.$$

In this way we construct the following objects.

An increasing sequence $W_1, W_2, \ldots$ of separable linear subspaces of V.

An increasing sequence $\{z_j^{(1)}\}, \{z_j^{(2)}\}, \ldots$ of countable sequences in V with

$$d_i(z_j^{(n)}), \quad i \geq 1, \quad j \geq 1,$$

dense in $D(W_n)$, for $n \geq 1$.

An increasing sequence of countable sets $\mathcal{D}_n$ in X^*, $\mathcal{D}_n$ consisting of all finite rational linear combinations of points from $D(\{z_j^{(n)}\})$.

A sequence $\mathcal{V}_2, \mathcal{V}_3, \ldots$ of countable subsets of V, with

$$V_{n+1} = \{v(n, d) : d \in D_n\},$$

where $v(n, d)$ is chosen for each d in $\mathcal{D}_n$ so that

$$\langle f - d,\ v(n,d) \rangle \geq \|f\|_V - d|_V\|_{V^*} - (1/n),$$

and $\|v(n,d)\| \leq 1, v(n,d) \in V$.

We also have

$$W_{n+1} = \overline{\mathrm{sp}}\{W_n \cup V_{n+1}\}$$

for $n \geq 1$.

We complete the construction of Y to be the norm closure

$$Y = \mathrm{cl}\, W$$

of the linear space W, which is not in general closed. Note that

$$Y = \mathrm{cl} \bigcup_{n=m}^{\infty} W_n$$

for each $m \geq 1$. Clearly Y is a separable subspace of V.

Since $Y \subset V$, we have $f|_Y \in Y^*$. Since Y is separable we can apply Step 2. This yields

$$f|_Y \in Y^* = \overline{\mathrm{sp}}\{D_Y(Y)\} = \overline{\mathrm{sp}}\{x^*|_Y : x^* \in D(Y)\}.$$

Let $\epsilon > 0$ be given. Choose $m \geq 2/\epsilon$. Now we can choose a point g in

$$\mathrm{sp}\{D_Y(Y)\}$$

with

$$\|f|_Y - g\|_{Y^*} < \epsilon/2.$$

Now g will be a finite linear combination of points in $D_Y(Y)$. So we can replace g by a finite rational linear combination h of the same points of $D_Y(Y)$, and still have

$$\|f|_Y - h\|_{Y^*} < \epsilon/2.$$

Since

$$Y = \mathrm{cl} \bigcup_{n=m}^{\infty} W_n$$

and the functions d_i are continuous, the point h can be replaced by the same finite rational linear combination j of points taken from $D_Y(\bigcup_{n=m}^{\infty} W_n)$, and still have

$$\|f|_Y - j|_Y\|_{Y^*} < \epsilon/2.$$

Since the sequences $\{d_i(z_j^{(n)})\}_{i \geq 1, j \geq 1}$ are dense in $D(W_n)$ for each n, we can replace the finite rational linear combination j by the same finite rational linear combination k of points in

$$\bigcup_{n=m}^{\infty} D(\{z_j^{(n)}\}),$$

with

$$\|f|_Y - k|_Y\|_{Y^*} < \epsilon/2.$$

Let ℓ be the largest integer, necessarily at least m, involved as an index n in the representation of k in terms of points from the sets $D(\{z_j^{(n)}\})$, $n \geq m$. Then k is a finite rational linear combination of points in

$$\bigcup_{n=m}^{\ell} D(\{z_j^{(n)}\}).$$

Since the sequences $\{z_j^{(n)}\}, n \geq 1$, are increasing, k is a finite rational linear combination of points from $D(\{z_j^{(\ell)}\})$ and so k is a point of $\mathcal{D}_\ell$. Hence there is a point $v(\ell, k)$ in $\mathcal{V}_{\ell+1}$ satisfying $\langle f - k, v(\ell, k) \rangle \geq \|f|_V - k|_V\|_{V^*} - (1/\ell)$ and $\|v(\ell, k)\| < 1$.

Now

$$\|f|_V - k|_V\|_{V^*} \leq (1/\ell) + \langle f - k, \ v(\ell, k) \rangle$$

$$= (1/\ell) + \langle f|_Y = k|_Y, \ v(\ell, k) \rangle$$

$$\leq (1/m) + \|f|_Y - k|_Y\|_{Y^*}$$

$$< (\epsilon/2) + (\epsilon/2) = \epsilon.$$

Recall that f being in V^* coincides with $f|_V$, and that $k|_V \in \overline{\mathrm{sp}} \, D|_V(V)$. Thus the distance of f from $\overline{\mathrm{sp}} \, D|_V(V)$ is at most ϵ. Since this holds for all $\epsilon > 0$ we have

$$f \in \overline{\mathrm{sp}}\{D|_V(V)\}.$$

Since f may be any point of V^*,

$$V^* = \overline{\mathrm{sp}}\{x^*|_V : x^* \in D(V)\}$$

as required.

Step 4. We remark that the set-valued map $D : X \longrightarrow 2^{X^*}$ introduced in Step 1 is a norm to weak lower semi-continuous map with $D(x)$ a countable set for every x in X.

Now Fabian [9], following but modifying work of John and Zizler [42, 43], has proved the following theorem.

Theorem (Fabian)   *Let X be a Banach space with a norm to weak lower semi-continuous multivalued mapping $D : X \longrightarrow 2^{X^*}$ such that D_x is a counta-ble set for every x in X and that*

$$\overline{\text{sp}}\{x^*|_V : x^* \in D(V)\} = V^*$$

for every subspace V of X.

Then, if μ is the first ordinal with cardinality that of dens X, there exists a nondecreasing "long sequence" $\{M_\alpha : \omega \leq \alpha \leq \mu\}$ of subspaces of X and a "long sequence" $\{P_\alpha : \omega \leq \alpha < \mu\}$ of linear projections on X^* forming a Projective Resolution of the Identity, the sequence $\{M_\alpha : \omega \leq \alpha \leq \mu\}$ satisfying the conditions:

(i) dens $M_\alpha \leq \overline{\alpha}$;

(ii) $\cup_{\beta<\alpha} M_{\beta+1}$ is dense in M_α;

(iii) the mapping $R_\alpha : P_\alpha X^* \to M_\alpha^*$ defined by $R_\alpha f = f|_{M_\alpha}$, $f \in P_\alpha X^*$, is a surjective isometry and $P_\alpha f = R_\alpha^{-1}(f|_{M_\alpha})$ for all $f \in X^*$.

By the initial remark of this step and the result of Step 3, when X is an Asplund space, X satisfies the hypothesis of Fabian's theorem. The conclusions of Fabian's theorem combined with known techniques shows that the dual of any Asplund space has an equivalent locally uniformly convex norm.

7.3 A PARTIAL CONVERSE TO THEOREM 5.4

This section is much easier to follow than Section 7.2. We first prove Theorem 7.1 stated in the introduction to this chapter. We then verify the following example showing that something like the proviso is needed in this theorem.

Example 7.1   *The attainment map from the space $\ell_1(\mathbb{N})$ to the unit ball B^* of the dual space $\ell_\infty(\mathbb{N})$ of $\ell_1(\mathbb{N})$, has a selector that is of the first Baire class as a map from $(\ell_1(\mathbb{N}), \text{norm})$ to $(\ell_\infty(\mathbb{N}), \text{norm})$, but (B^*, weak^*) is not fragmented by the norm.*

The proof of Theorem 7.1 follows (but elaborates) the first part of the proof Theorem 26 of a paper by Jayne, Orihuela, Pallarés and Vera [41], but avoids the lacuna in their proof by use of the proviso written into our restatement of the theorem and by appeal to Lemma 7.2.

Proof of Theorem 7.1. Let K^* be a nonempty weak* compact set in the dual X^* of a Banach space X that contains no isomorphic copy of ℓ_1. Suppose further the attainment map F from X to K^* has a selector f that is of the first Baire class as a map from (X, norm) to (X^*, norm).

In order to show that (K^*, weak^*) is fragmented by the norm of X^*, we seek to apply a result of Namioka [60, Theorem 3.4], which generalizes an earlier result of Stegall [75]. Namioka considers a weak* compact set in the dual X of a Banach space, If A is a bounded subset of X he introduces a pseudo norm n_A for X^* defined by

$$n_A(x^*) = \sup\{|\langle x, x^* \rangle| | x \in A\}.$$

He proves, in particular, that *if K^* is separable for n_A for each bounded countable subset A of X, then $(K^*, weak)$ is fragmented by the norm in X^*.* In this result, it is clearly sufficient to confine attention to countable subsets A of the unit ball of X. Then

$$n_A(x^*) = \sup\{|\langle x, x^* \rangle| : x \in A\}$$

$$\leq \sup\{|\langle x, x^* \rangle| : \|x\| < 1\} = \|x^*\|.$$

Now for any countable set A contained in the unit ball of X we consider the closed linear span S of A. S is a closed separable subspace of X. By the Hahn–Banach theorem, every linear functional defined on S can be extended, without change of norm to a linear functional on X. It follows that the dual space S^* of S can be identified with the functions on S obtained by restriction to S of the linear functionals on X, the norm of such a function g on S being taken to be

$$\|g\| = \sup\{g(x) : x \in S, \|x\| \leq 1\}.$$

We use p to denote the restriction map $p : X^* \to S^*$, with $p(x^*)$ taken to be the restriction of x^* to S, for each x^* in X^*. Clearly p is a linear map with norm 1.

We verify that p is a continuous map from (X^*, weak^*) to (S^*, weak^*) A typical basic weak* open neighborhood of the origin of S^* is of the form

$$\{s^* \in S^* : \langle s_i, s^* \rangle < 1 \text{ for } 1 \leq i \leq n\}$$

for points $s_1, s_2, \ldots, s_n$ in S. The inverse image under p of this weak* open neighborhood is

$$\{x^* \in X^* : \langle s_i, x^* \rangle < 1 \text{ for } 1 \leq i \leq n\}$$

for the same points $s_1, s_2, \ldots, s_n$ in S. Hence p is continuous as a map from (X^*, weak^*) to (S^*, weak^*). Since K^* is weak* compact in X^*, it follows that $p(K^*)$ is weak* compact in S^*.

Recall that we are supposing that f is a Baire class 1 selector for the attainment map for K^*. So

$$\langle s, f(s) \rangle = \sup\{\langle s, k^* \rangle : k^* \in K^*\}$$

for each s in S. Thus, for s in S and k^* in K^*,

$$\langle s, p(k^*) \rangle = \langle s, k^* \rangle \leq \langle s, f(s) \rangle = \langle s, p(f(s)) \rangle$$

and $p(f(s)) \in p(K^*)$. Hence, for s in S and s^* in $p(K^*)$,

$$\langle s, s^* \rangle \leq \langle s, p(f(s)) \rangle \quad \text{and} \quad p(f(s)) \in p(K^*).$$

This means that the set

$$\{p(f(s)) : s \in S\}$$

is a boundary for $p(K^*)$ in S^* and $p \circ f$ is a selector for the attainment web from S to $p(K^*)$. Since p is continuous and f is of the first Baire class, $p \circ f$ is also of the first Baire class from (S, norm) to $(S^* \text{ norm})$.

Since we are assuming that X contains no isomorphic copy of $\ell_1(\mathbb{N})$, the subspace S can contain no such copy. Now S, S^*, $p(K^*)$ and $p \circ f$ satisfy the conditions of Lemma 7.2. Hence, by that lemma, $p(K^*)$ contains a countable norm dense subset, D_o say. For each point d_o in D_o pick a point d_1 in K^* with $p(d_1) = d_o$, and take D_1 to be the set chosen in this way. Then D_1 is a countable set in K^* that is dense with respect to the pseudo norm q on X^* defined by

$$q(x^*) = \|p(x^*)\|.$$

Since A is contained in the unit ball of S,

$$n_A(x^*) = \sup\{|\langle x, x^*\rangle| : x \in A\}$$

$$\leq \sup\{|\langle x, x^*\rangle| : x \in S, \|x\| \leq 1\}$$

$$= \|p(x^*)\| = q(x^*).$$

Thus D_1 is a countable set in K^* that is dense in K^* with respect to the pseudo norm n_A.

Applying Namioka's result we conclude that (K^*, weak^*) is fragmented in X^* by the norm.

Verification of Example 7.1   Regard $\ell_\infty(\mathbb{N})$ as the dual of $\ell_1(\mathbb{N})$ Then B^* is a convex weak* compact set in $\ell_\infty(\mathbb{N})$.

The attainment map $F : \ell_\infty(\mathbb{N}) \to B^*$ is defined by

$$F(x) = \{x^* : \langle x, x^*\rangle = \sup\{\langle x, y^*\rangle : \|y^*\| \leq 1\}$$

$$F(x) = \{x^* : \langle x, x^*\rangle = \sup\{\langle x, y^*\rangle : \|y^*\| \leq 1 \text{ and } \|x^*\| \leq 1\}.$$

If $x = \{x_1, x_2, \ldots\}$ in $\ell_\infty(\mathbb{N})$ and $y^* = \{y_1^*, y_2^*, \ldots\}$ in $\ell_\infty(\mathbb{N})$, then

$$\sup\{\langle x, y^*\rangle : \|y^*\| \leq 1\} = \sup\left\{\sum_{i=1}^\infty x_i y_i^* : |y_i^*| \leq 1, \ i \geq 1\right\}$$

$$= \sum_{i=1}^\infty |x_i|,$$

and this supremum is attained just when

$$y_i^* = \text{sign } x_1, \quad \text{if } x_1 \neq 0,$$

$$y_i^* \in [-1, 1], \quad \text{if } x_i = 0.$$

Thus

$$F(x) = \{y^* : y_i^* = \text{sign } x_i \text{ if } x_i \neq 0, y_i^* \in [-1, 1] \text{ if } x_i = 0\}.$$

A selector for this attainment map is the function $f: \ell_\infty(\mathbb{N}) \to B^*$ defined by

$$f(x) = y^*$$

with

$$y_i^* = \text{sign } x_i \quad \text{if } x_1 \neq 0$$

$$y_i^* = 0 \qquad\quad \text{if } x_i = 0.$$

Now, for each $n \geq 1$, take

$$f^{(n)}(x) = y^{*(n)},$$

with

$$y^{*(n)} = |x_i| \tanh(nx_i), \quad i \geq 1.$$

Then for each fixed x in $\ell_\infty(\mathbb{N})$ the sum $\sum |x_i|$ converges and

$$\lim_{n \to \infty} f^{(n)}(x) = f(x) \quad \text{in } \ell_\infty(\mathbb{N}).$$

Further, for each fixed $n \geq 1$, the map $f^{(n)} : \ell_\infty(\mathbb{N}) \to \ell_\infty(\mathbb{N})$ is continuous. Thus the attainment map has a Baire first class selector. However the set $(B^*,$ weak$^*)$ is not even σ-fragmented by the norm in $\ell_\infty(\mathbb{N})$.

7.4 REMARKS

1. For the convenience of the masochistic reader we quote the statement of Simons' multipurpose lemma that we have not used in the proof of Lemma 7.1. We recall the notation he uses. When X is a nonempty set and $f \in \ell_\infty(X)$ write $S(f) = \sup f(X)$ and $\|f\| = \sup |f(X)|$.

   Lemma (Simons) *We suppose that for all $n \geq 1$, $f_n \in \ell_\infty(X)$ and $\sup_{n \geq 1} \|f_n\| < \infty$. We suppose further that $Y \subset X$ and that, whenever $\lambda_n \geq 0$ and $\sum_{n \geq 1} \lambda_n = 1$, there exists $y \in Y$ such that $\sum_{n \geq 1} \lambda_n f_n(y) = S(\sum_{n \geq 1} \lambda_n f_n)$. Then*

   $$\sup_{y \in Y} \lim \sup_{n \to \infty} f_n(y) \geq \inf S(\text{conv}\{f_n : n \geq 1\}).$$

2. The lacuna in the proof of Theorem 26 of the paper by Jayne, Orihuela, Pallarés and Vera arose since they applied Theorem I.2 of Godefroy [15] without verifying that the conditions of this theorem were satisfied.

3. We conjecture that *Lemma 7.1 holds without the condition that the Banach space X be separable.* Perhaps this can be established by use of Fabian and Godefroy's method of reduction to the separable case. An alternative approach might be to modify the result of Haydon [24, Theorem 3.1]. Haydon proves that *if a Banach space contains no isomorphic copy of ℓ_1, then every weak* compact convex subset of X^* is the norm closed convex hull of its extreme points.*

Bibliography

[1] E. Asplund, Fréchet differentiability of convex functions, *Acta Math.* **121** (1968) 31–47.

[2] S. Banach, Über analytisch darstellbare, Operationen in abstrakten Räumen, *Fund. Math.* **17** (1931) 283–295.

[3] R. G. Bartle and L. M. Graves, Mappings between function spaces, *Trans. Amer. Math. Soc.* **72** (1952) 400–413.

[4] E. Borel, Quelques remarques sur les principles de la theorie des ensembles, *Math. Ann.* **60** (1905) 194–195.

[5] S. Braun, Sur l'uniformisation des ensembles fermes, *Fund. Math.* **28** (1937) 214–218.

[6] G. Choquet, Convergences, *Ann. Univ. Grenoble* **23** (1947–78) 57–112.

[7] R. Deville, G. Godefroy and V. Zizler, *Smoothness and Renormings in Banach Spaces*, Longman Scientific Technical, Harlow, 1993.

[8] R. Engelking, *General Topology*, PWN, Warsaw, 1977.

[9] M. Fabian, On projectional resolution of identity on the duals of certain Banach spaces, *Bull. Austral. Math. Soc.* **35** (1987) 363–371.

[10] M. Fabian and G. Godefroy, The dual of every Asplund space admits a prospective resolution of the identity, *Stud. Math.* **91** (1988) 141–151.

[11] H. Flanders, The Steiner point of a closed hyperspace, *Mathematika* **13** (1966) 181–188.

[12] K. Floret, *Weakly Compact Sets*, Lecture Notes in Mathematics, No. 801, Springer, Berlin, 1980.

[13] M. Fosgerau, When are Borel function Baire functions? *Fund. Math.*, **143** (1993) 137–152.

[14] N. Ghoussoub, B. Maurey and W. Schachermayer, Slicings, selections and their applications, *Canad. J. Math.* **44** (1992) 483–504.

[15] G. Godefroy, Boundaries of a convex set and interpolation sets, *Math. Ann.* **277** (1987) 173–184.

[16] B. Grünbaum, *Convex Polytopes*, John Wiley & Sons, London, 1976.

[17] J. Hadamard, *Œuvres de Jacques Hadamard*, Centre National de la Recherche Scientifique, Paris, 1968.

[18] J. Hadamard, R. Baire, H. Lebesgue and E. Borel, Cinq letters sur la theorie des ensembles, *Bull. Soc. Math. France* **33** (1904) 261–273.

[19] R. W. Hansell, Borel measurable mappings for nonseparable metric spaces, *Trans. Am. Math. Soc.* **161** (1971) 145–169.

[20] R. W. Hansell, Extended Bochner measurable selectors, *Math. Ann.* **277** (1987) 79–94.

[21] R. W. Hansell, J. E. Jayne, I. Labuda and C. A. Rogers, Boundaries of and selectors for upper semi-continuous set-valued functions, *Math. Z.* **189** (1985) 297–318.

[22] R. W. Hansell, J. E. Jayne and M. Talagrand, First class selectors for weakly upper semi-continuous multivalued maps in Banach spaces, *J. Reine Angew. Math.* **361** (1985) 201–220 and **362** (1986) 219–220.

[23] G. H. Hardy and E. M. Wright, *The Theory of Numbers*, 2nd edition, Clarendon Press, Oxford, 1945.

[24] R. G. Haydon, Some more characterizations of Banach spaces containing ℓ_1, *Math. Proc. Camb. Philos. Soc.* **80** (1978) 269–276.

[25] R. G. Haydon, J. E. Jayne, I. Namioka and C. A. Rogers, Continuous functions on totally ordered spaces that are compact in their order topologies, *J. Funct. Anal.* **178** (2000) 23–63.

[26] J. E. Jayne and C. A. Rogers, Fonctions multivoques semi-continues supérieurement, *C.R. Acad. Sci. Paris Sér. I* **293** (1981) 429–430.

[27] J. E. Jayne and C. A. Rogers, Upper semi-continuous set-valued functions, *Acta Math.* **149** (1982) 87–125.

[28] J. E. Jayne and C. A. Rogers, Sélections boréliennes de multi-applications semi-continues supérieurement, *C.R. Acad. Sci. Paris Sér. I* **299** (1984) 125–128.

[29] J. E. Jayne and C. A. Rogers, Borel selectors for upper semi-continuous multi-valued functions, *J. Funct. Anal.* **56** (1984) 279–299.

[30] J. E. Jayne and C. A. Rogers, Borel selectors for upper semi-continuous multi-valued functions, *Mathematika* **32** (1985) 324–337.

[31] J. E. Jayne and C. A. Rogers, Borel selectors for upper semi-continuous set-valued maps, *Acta Math.* **155** (1985) 41–79.

[32] J. E. Jayne and C. A. Rogers, Upper semi-continuous set-valued functions, *Acta Math.* **155** (1985) 149–152.

[33] J. E. Jayne and C. A. Rogers, Radon measures on Banach spaces with their weak topologies, *Serdica Math. J.* **21** (1995) 283–334.

[34] J. E. Jayne, I. Namioka and C. A. Rogers, Norm fragmented weak* compact sets, *Collect. Math.* **41** (1990) 133–163.

[35] J. E. Jayne, I. Namioka and C. A. Rogers, σ-fragmentable Banach spaces, *Mathematika* **39** (1992) 164–188 and 197–215.

[36] J. E. Jayne, I. Namioka and C. A. Rogers, Topological properties of Banach spaces, *Proc. London Math. Soc.* (3) **66** (1993) 651–672.

[37] J. E. Jayne, I. Namioka and C. A. Rogers, Fragmentability and σ-fragmentability, *Fund. Math.* **143** (1993) 207–220.

[38] J. E. Jayne, I. Namioka and C. A. Rogers, σ-fragmentable Banach spaces II, *Stud. Math.* **111** (1994) 69–80.

[39] J. E. Jayne, I. Namioka and C. A. Rogers, Continuous functions on compact totally ordered spaces, *J. Funct. Anal.* **134** (1995) 261–280.

[40] J. E. Jayne, I. Namioka and C. A. Rogers, Continuous functions on products of compact Hausdorff spaces, *Mathematika* **46** (1999) 323–330.

[41] J. E. Jayne, J. Orihuela, A. J. Pallarés and G. Vera, σ-fragmentability of multivalued maps and selection theorems, *J. Funct. Anal.* **117** (1993) 243–273.

[42] K. John and V. Zizler, Duals of Banach spaces which admit nontrivial smooth functions, *Bull. Austral. Math. Soc.* **11** (1974) 161–166.

[43] K. John and V. Zizler, Smoothness and its equivalents in weakly compactly generated Banach spaces, *J. Funct. Anal.* **15** (1974) 1–11.

[44] J. L. Kelley and I. Namioka, *Linear Topological Spaces*, Springer, Berlin, 1963.

[45] P. S. Kenderov, Dense strong continuity of pointwise continuous mappings, *Pacific J. Math.* **89** (1980) 111–130.

[46] M. Kondô, Sur l'uniformisation des complémentariness analytiques et les ensembles projectifs de la seconde class, *Japan J. Math.* **15** (1939) 194–230.

[47] K. Kuratowski, *General Topology*, Vol. I, Academic Press, New York, 1966.

[48] K. Kuratowski and C. Ryll-Nardzewski, *Bull. Acad. Polon. Sci.* **13** (1965) 397–403.

[49] N. Lusin, Sur les ensembles analytiques, *Fund. Math.* **10** (1924) 1–95.

[50] N. Lusin, Sur les fonctions implicité à une infinite dénumbrable des valeurs, *C.R. Acad. Sci., Paris* **189** (1929) 313–316.

[51] N. Lusin, *Les Ensembles Analytiques*, Gauthier-Villars, Paris, 1930.

[52] N. Lusin and P. S. Novikov, Choise éffectif d'un point dans un complémentaire analytic arbitraire, donne pas un crible, *Fund. Math.* **25** (1935) 559–560.

[53] D. A. Martin and A. S. Kechris, Infinite games and effective descriptive set theory, in: C. A. Rogers, J. E. Jayne, C. Dellacherie, F. Topsøe, J. Hoffmann-Jørgensen, D. A. Martin, A. S. Kechris and A. H. Stone (editors), *Analytic Sets*, Academic Press, London, 1980, pp. 404–470.

[54] E. Michael, Continuous selections I, *Ann. Math.* (2) **63** (1956) 361–382.

[55] E. Michael, Continuous selections II, *Ann. Math.* (2) **64** (1956) 562–580.

[56] E. Michael, Continuous selections III, *Ann. Math.* (2) **65** (1957) 375–390.

[57] E. Michael, Selected selection theorems, *Amer. Math. Month.* **63** (1956) 233–238.

[58] G. J. Minty, On the monotonicity of the gradient of a convex function, *Pacific J. Math.* **14** (1964) 243–247.

[59] D. Montgomery, Non-separable metric spaces, *Fund. Math.* **25** (1935) 527–533.

[60] I. Namioka, Radon–Nikodým compact spaces and fragmentability, *Mathematika* **34** (1987) 258–281.

[61] I. Namioka and R. R. Phelps, Banach spaces which are Asplund spaces, *Duke Math. J.* **42** (1975) 735–750.

[62] I. Namioka and R. Pol, σ-fragmentability and analyticity, *Mathematika* **43** (1996) 172–181.

[63] P. Novikoff, Sur les fonctions implicites mesurables B, *Fund. Math.* **17** (1931) 8–25.

[64] L. E. Odell and H. P. Rosenthall, A double-dual characterization of separable Banach spaces containing ℓ^1, *Israel J. Math.* **20** (1975) 375–383.

[65] J. Pach and C. A. Rogers, Partly convex Peano curves, *Bull. London Math. Soc.* **15** (1983) 321–328.

[66] R. R. Phelps, *Convex Functions, Monotone Operators and Differentiability*, Lecture Notes in Mathematics, No 1364, Springer, Berlin, 1989.

[67] C. A. Rogers and R. C. Willmott, On the uniformization of sets in topological spaces, *Acta Math.* **120** (1968) 1–52.

[68] C. A. Rogers, J. E. Jayne, C. Dellacherie, F. Topsøe, J. Hoffmann-Jørgensen, D. A. Martin, A. S. Kechris and A. H. Stone (editors), *Analytic Sets*, Academic Press, London, 1980.

[69] W. Rudin, *Functional Analysis*, 2nd edition, McGraw-Hill, New York, 1991, 424 pp.

[70] Y. Sampei, On the uniformization of the complement of analytic set, *Comment. Math. Univ. St. Paul* **10** (1960) 54–62.

[71] W. Sierpiński, Sur l'uniformisalion des ensembles mesurables (B), *Fund. Math.* **16** (1930) 136–139.

[72] S. Simons, A convergence theorem with boundary, *Pacific J. Math.* **40** (1972) 703–708.

[73] M. Sion, On uniformazation of sets in topological spaces, *Trans. Am. Math. Soc.* **96** (1960) 237–245.

[74] V. V. Srivatsa, Baire class 1 selectors for upper semi-continuous set-valued maps, *Trans. Amer. Math. Soc.* **337** (1993) 609–624.

[75] C. Stegall, The Radon–Nikodým property in conjugate Banach spaces, *Trans. Amer. Math. Soc.* **206** (1975) 213–223.

[76] A. H. Stone, Paracompactness and product spaces, *Bull. Amer. Math. Soc.* **54** (1948) 977–982.

[77] A. H. Stone, Non-separable Borel sets, *Dissertationes. Math. (Rozprawy Mat.)* **18** (1962) 1–40.

[78] A. H. Stone, Non-separable Borel sets II, *Gen. Top. Appl.* **2** (1972) 249–270.

[79] Y. Suzuki, On the uniformization principle, in: *Proceedings of the Symposium on the Foundations of Mathematics*, Katada (Japan), 1962, 137–144.

[80] M. Talagrand, Renormage de quelques $C(K)$, *Israel J. Math.* **54** (1986) 327–334.

[81] E. Zermelo, Beweis, das jede Menge wohlgeordnet werden kann, *Math. Ann.* **59** (1904) 514–516.

Index